Revised Print

Ethnobotany
A Modern Perspective

Rani Vajravelu, Ph.D

Kendall Hunt
publishing company

With due respects and appreciation, I dedicate this work to

Dr. Henry O. Whittier

Retired Professor of Biology and Former Director of the Arboretum
University of Central Florida

Cover image courtesy of Rani Vajravelu

Kendall Hunt
publishing company
www.kendallhunt.com
Send all inquiries to:
4050 Westmark Drive
Dubuque, IA 52004-1840

BRIEF CONTENTS

CONTENTS

Chapter 10   Plant Narcotics and Poisons   173

Chapter 11   Plant Fibers and Wood Products   187

PREFACE

Ethnobotany: A Modern Perspective focuses on the indigenous plants and their impact on modern society. The book discusses two hundred and sixty plants that originated in different parts of the world and how they relate to human lives every day. It is for the current generation of students who may need a course in Ethnobotany, Plants and Humanities, Plants and People, Cultural Botany, Culinary Botany, or similar courses that focus on plants in human life. The goal is to provide a comprehensive, up-to-date, affordable, college-level textbook without losing out on the allure of ethnobotany.

The main content of the book is built on the indigenous plants used around the world as food, beverage, medicine, narcotics, shelter, and clothing, along with their applications elsewhere. Line drawings, appropriate terminology, background botanical information, examples, tables or maps or, often both, along with review questions supplement each chapter. The presented material is at the major's level. However, one can adopt this book in a non-major course as the technical details level with well-known examples and labeled drawings. The text maintains a delicate balance throughout between common usage and scientific information. To satisfy all levels of students, the textbook refers to each plant by its common name, its scientific name, and the family to which it belongs. The professor can easily disregard the scientific names if he or she deems it necessary for the nature of the course. Each chapter starts with an introduction that is relevant to the subject matter and ties in with real world applications. The book includes a separate compendium of original color photographs of the plants and their products in a separate website.

Pedagogical Organization

This textbook includes eleven chapters. Some of the chapters further divide into two or more sections, making it easy for the professors to pick and choose the areas they want to cover. The content of the book is suitable for one semester or one quarter courses at the junior level. The inclusion of necessary background information throughout the book to help the reader understand the chapter content eliminates the need for a prerequisite basic botany course.

CHAPTER 1: Introduces **ethnobotany** as a separate discipline and provides highlights about the **applications** and importance of this field. It includes a discussion on ethnobotany and economic botany, the pioneers in the field, and the training needed to become an ethnobotanist. Such information is vital to a text of this scope. The chapter tries to provide an awareness of how our lives depend on plants, most often, the introduced crops. It presents a list of the most widely cultivated plant crops and their places of origin along with a map. Another aspect is an outline of **plant nutrients** found in the edible plants discussed throughout the book.

CHAPTER 2: Since basic knowledge about botany is essential to understand and enjoy the text, a single chapter, ***Botany in a Nutshell***, highlights and summarizes all the necessary botanical information. Original line drawings and color photographs explain the information to anyone who is new to botany and form a good connection for the already trained botanist.

CHAPTERS 3 AND 4: These two chapters discuss **vegetables** and **fruits** separately. They include a minimum of 42 families and about 103 vegetables and fruits. Vegetables and fruits are logically categorized as cruciferous vegetables, citrus fruits, and so on. The chapters discuss each plant's indigenous use, place of origin, and current distribution. Wherever applicable, short descriptions of plants along with additional historical connections are added. The chapters also discuss the plant's impact on human society and any commercial products available from it. Whenever suitable, the chapters add information on available nutrients, mention **superfruits**, and discuss popular exotic products. To remove longtime confusion about the definitions of fruit and vegetables and fruits or nuts, separate topics on *fruit or vegetable* and *fruits that go nuts* are included with botanical explanations and well known examples. Even though the chapters discuss most widely known plants from around the world, care was taken to include even lesser known yet locally popular plants, like mangosteen and malay apples.

CHAPTER 5: **Grains** and **legumes** are combined in one chapter and discussed in two separate sections. They include a total of 24 plants that belong to three families. Since each group is unique on its own, the beginning of each section presents the necessary botanical background with highlights such as *from the field to the pantry* on grain selections and *"grains" that are not real grains* with reference to quinoa and buckwheat. Terms like whole grain, essential amino acids, soluble fiber, and many more in this category have become the everyday norm for modern society, and students are no exception. In order to help them understand and make educated choices, the chapter includes scientifically supported information. One would find the information on *dhal and lentil* and the Whole Grains Council interesting and applicable in day to day life.

CHAPTERS 6 AND 7: **Beverages** derived from 14 plant sources are the basis of discussion in Chapter 6; **vegetable oils** from 19 sources are the focus in Chapter 7. Both chapters discuss plant substances rather than entire plant parts. Such substances have stimulating effects and/or health connections. Each of these chapters is built on its own strength. Detailed information for important beverage sources includes the legends behind the discovery and first uses and impact on society. For example, students would find it interesting to learn about the original discovery of tea in the Orient; its role in Chinese and Japanese traditions; processing methods; chemical compounds; its spread throughout the world and its impact on the colonists and history of North America; the compounds that give the flavor of the tea; the differences between black tea and the exotic **chai**; and the most recent *Brassica* tea and its health benefits. Similarly, Chapter 7 deals with indigenous uses of oil plants and health connections. It includes **rice bran oil** and **neem oil** along with their indigenous connections. Students will be able to make informed healthy choices after going through these chapters.

CHAPTER 8: Focuses entirely on **spices and herbs** that enhance the flavor of our food. The chapter categorizes a total of 38 aromatic plants as either spices or herbs. Since the aromatic plants have encouraged voyages across the oceans, conquering lands, and starting wars, it was appropriate to include a brief account about the **history of spices and herbs** at the beginning of the chapter. In addition to providing ethnobotanical information, the chapter outlines the culinary uses for each plant. It also includes spices newer to the Western world, such as mango ginger, asafetida, and fenugreek. Discussion about wasabi and capers would appeal to most students. The picture CD includes a display of 22 different aromatic plants, individually labeled. When talking about spices and herbs there is always a question about **curry powder**. To remove any confusion regarding curry powder, the chapter includes a separate discussion sure to draw anyone towards curry.

CHAPTER 9 AND 10: These two chapters focus on the medicinal, narcotic, and poisonous effects of about 42 new plants not discussed in the previous chapters. Non-food plants are discussed for the first time in the book with the same level of attention given to edible plants so far. Both chapters classify active chemical compounds in a way that is easy on students who have limited training in chemistry.

The **history of herbal medicine** is an important aspect of Chapter 9. Chapter 10 begins with the discussion of the addictive effects of plant narcotics on the human body and mind followed by a list of **misused plants** and their discovery and spread throughout the world. The discussion expands to the first use, cultural impact, religious connections, and names and effects of each chemical compound. The chapter includes two world maps, one to show the ten most important medicinal plants in the world and the other to show the spread of marijuana from the east to the west. Summary tables outline the included plants for students to review.

CHAPTER 11: Discusses the **fiber and wood** derived from plants, including everything from the first paper to the most modern bamboo floor. It is divided into two separate sections on fiber and wood with a total of 27 plant sources. To understand the nature of fiber and wood, the chapter briefly presents necessary botanical information further supported by illustrations. It presents all the commercial products, including, but not limited to, the dollar bills, tea bags, and wood veneers. Students will develop the same appreciation for these plants as they did in previous chapters. Overall, this chapter will help students to develop an appreciation for the plants in general and also for ethnobotany.

Additional Features

The **summary tables** included in most chapters comprise the information for many plants that belong to a single category. Students can compare and review the content in one area.

World maps show the distribution, origin, or spread of plants in discussion across the world.

The **review questions** at the end of each chapter include ten each of free response type, multiple choice, and true or false questions. The design of the review questions is to encourage students to develop a deep understanding of the content and to help them retain the information. Use them as sample questions to prepare for the test. Due to the vast amount of information, any number of questions can be generated following the given format.

A new and improved image library is under construction which will include the images from the picture CD of the previous edition and many more. The photographs include rare exotic trees like durian, neem, chocolate, tea, mahogany and many more taken from at least five different countries and eight different botanical institutions including the Forest Research Institute, Malaysia; the Royal Botanic Gardens, Kew, U.K.; and Butchart Gardens in Vancouver, Canada. Many of the photographs include multiple items; some are categorized by the family or by ethnobotanical uses, such as oil yielding plants or fiber sources. Each photo has a name that connects with the information given in the text; most have clear labels for students to help them study on their own. Please use the following URL to access the image library http://webcom3.grtxle.com/ancillary.

References

A book of this nature must be built upon the knowledge of others. I would like to acknowledge the listed references at the end of the book for providing a wealth of information that framed most of this textbook. Any deviation or errors of interpretation are mine and do not reflect on the original sources. A single textbook with limitations on size cannot possibly accommodate the vast amount of information available for the 92 families included in this book. The references at the end of the book offer several wonderful sources that expand on each of the topics. Anyone who would like to read further on any given topic should find them useful. A separate list of references is provided for Appendix 1. We credit the listed sources for providing the data.

Acknowledgement

This textbook takes into account the feedback received from my past students of ethnobotany course who have tactfully indicated what they would like in an ethnobotany textbook. I thank them for influencing me indirectly into writing a textbook of my own. I thank Tina Richards for providing valuable feedback in improving the content of this edition. I acknowledge Kimberly Giulvezan for compiling the material included in Appendix 1. Preparing a book of this nature was a monumental effort on my part considering the short timeframe that was available. I could not have accomplished this without the support of my family who helped me stay on the track and offered help whenever needed. It is my sincere hope that this book will satisfy the intellectual needs of the present generation of students. I look forward to receiving comments and criticism from anyone who uses this book, which future works will gladly accommodate.

Rani Vajravelu
Department of Biology
University of Central Florida
Orlando, Florida
e mail: rani.vajravelu@ucf.edu
Phone: 407 823 0990

LIST OF PHOTOGRAPHS

Original photographs of the following are included in the textbook website. Please use the following URL to access the image library http://webcom3.grtxle.com/ancillary. Photographs were taken by the author unless marked otherwise. The website will include a new image library which is under construction.

CHAPTER 1

1. Plants in Life 1
2. Plants in Life 2*
3. Harvard University Herbarium
4. Voucher Specimen

CHAPTER 2

1. Angiosperms
2. Pine
3. Fern
4. Moss
5. Flowers
6. Parts of a Flower—Dicot
7. Parts of a Flower—Monocot
8. Bougainvillaea Plant
9. Unisexual Flowers
10. Latex
11. Linnaeus Statue

CHAPTER 3

1. Vegetable Display
2. Cruciferous Vegetables
3. Apiaceae Vegetables
4. Starchy Vegetables
5. Sweet Potato
6. Taro Plant
7. Cassava Plant
8. *Allium* Vegetables
9. Onion Parts

* Vajravelu, Gopi
** Whittier, Henry
*** Used under license from Shuttercock, Inc.

CHAPTER 1
ETHNOBOTANY AND
ITS APPLICATIONS

Please recall all of the plants and their products that you have encountered since this morning. Start with your breakfast and think about linens, clothes, cosmetics, paper, and pencils and then continue the list. For every plant product that you have used, come up with another non-plant alternative. How would life feel without plants around you? By now, you might have come to the realization that plants are an integral part of your life and that of others as well. Such a relationship between plants and people is nothing new; it has existed since the Stone Age. Early humans searched for grains and nuts to supplant their hunting and gathering life style. That search for plants is still going on today to satisfy greater needs that developed over time. Several areas of science aim at fulfilling such needs by focusing on the growth, development, cultivation, yield, and manipulation of plants. Two branches among them directly deal with plants and people and are interrelated in many ways.

Ethnobotany and Economic Botany

Ethnobotany is the study of traditional plant uses by indigenous people. The word derives from "*ethno*" for culture and "*botany*" for the study of plants. John William Harshberger, a Florida botanist, used the term "ethnobotany" for the first time in 1895. Even though the idea of ethnobotany is ancient, as an academic science it is a relatively new branch of botany that has seen an increased awareness since the beginning of the twentieth century. It is emerging as an interdisciplinary science that incorporates training in plant taxonomy and cultural anthropology. Ethnobotanical research utilizes information from other fields such as archaeology, biochemistry, history, linguistics, religion, and mythology.

Economic botany is the study of the economic importance of plants. This field focuses on the application of botanical knowledge to the well-being of human populations, whereas, ethnobotany focuses on the indigenous uses of plants among native people. Economic botany is concerned with improving crops like corn for better yield, disease resistance, and biodiesel, whereas ethnobotany talks about how Native Indians used corn in their diet and culture. Our textbook, *Ethnobotany: A Modern Perspective,* incorporates the indigenous uses as well as the modern applications of plants in human society.

Plants and people have had a close association since the beginning of human civilization. Ancient people investigated plants for their primary needs, such as food to survive, medicine to heal, and fuel to light up; and naturally they have learned their uses. Over time they learned to manipulate plants to suit their needs and accumulated practical knowledge. Such native wisdom remained within a community and became transferred from one generation to another. Since people hardly migrated too far away from their native lands, each tribe

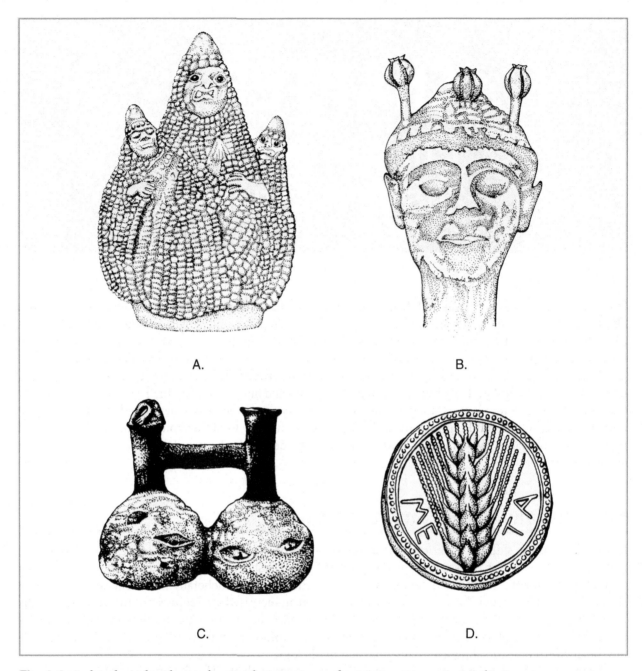

Fig. 1.1 Archaeological evidences showing the importance of certain crops in ancient civilizations

    A. A vase from Moche civilization showing three corn gods. Mochican people cultivated corn in Peru between the first and the third centuries.

    B. A stone sculpture adorned with poppy capsules obtained from the island of Crete in the Mediterranean Sea.

    C. A potato pot used by Chimu culture in Peru. Illustration based on an artifact in the British Museum.

    D. A Greek coin showing Barley, which was an important item of trade in ancient times.

contained its own information. Later on, as the needs increased, people started to move around looking for additional reliable plant sources either for personal use or for financial gain. In the course of this search they exchanged their knowledge with neighboring tribes. Plants formed a connection among people who utilized them in a similar fashion; that led to a sense of community with similar culture and traditions. Such associations created special symbolic meaning for the plants in religion and worship. As technology advanced, people learned to cross the borders, to explore the world, and to discover efficient ways for growing and processing plants. More often, along with their migration they brought their favorite plants and planted them in new places. These plants slowly became incorporated into the new lands. For instance, tomatoes and corn are integral parts of everyone's life in the world even though Mexican tribes first discovered and used these two plants.

Ethnobotanists

Since ethnobotany is a multidisciplinary science, it incorporates the knowledge of trained botanists who work closely with native people and experts of other fields such as anthropology and biochemistry. Such trained people are the **ethnobotanists** who explore the relationship between the plants and indigenous people; research the potential uses of plants; and document the data in a methodical and scientific way. Ethnobotanists spend their time living with native people of different cultures, acquiring their native wisdom, and sharing their knowledge with the rest of the world. One way to do this is to publish their discoveries. One of the first such published books was by Leopold Glueck in 1896 on the ethnobotany of medicinal plants used by the rural people of Bosnia.

The revival of ethnobotany began in the 1900s, where it became transformed from a mere compilation of information on plant uses to the cultural significance of the lives of rural people. One notable contribution was by **Richard Evans Schultes** who incorporated ethnobotany into an academic curriculum. Schultes was born in 1915 in the United States. He was a plant explorer, ethnobotanist, and conservationist. His focus was the use of hallucinogenic plants by Native Americans, particularly the people of Mexico and the Amazon. Schultes spent several years in the Amazon collecting over 24,000 specimens, out of which 300 were new species. One of his ethnobotanical discoveries was the source of curare poison used by indigenous people. Now we know that plant is *Chondrodendron,* a deadly poisonous plant used as a muscle relaxant in modern medicine. He was known as the world's authority on medicinal, narcotic, and hallucinogenic plants of the New World. His research has benefited the medical field by his contribution of about 2,000 plants used by indigenous people. His books, *The Healing Forest: Medicinal and Toxic Plants*

Fig. 1.2 Dr. R. E. Schultes
Courtesy of the Economic Botany Archives of Oakes Ames. Harvard University. Cambridge, MA

Fig. 1.3a Mark Plotkin
(Amazon Conservation Team. Used by permission.)

of the Northwest Amazonia coauthored with Robert Raffauf, and *The Plants of the Gods* coauthored with Albert Hofmann, were some of his great contributions. Albert Hofmann was the chemist who discovered LSD. For all of Schultes' contributions to ethnobotany, H.R.H. Prince Philip named him the "Father of Ethnobotany." Schultes has continued to work at Harvard University, first as the curator and later as a professor, where he inspired students with his personal experience acquired in the forests of Amazon. He was also the Director of the Botanical Museum where he has deposited most of his ethnobotanical collections. Richard Schultes died in 2001 after leading many of his students into ethnobotany.

One such student was **Edmund Wade Davis** (born in 1953), who also explored the Amazon and the Andes mountains and made 6,000 botanical collections. One of his studies was about the Amazonian tea made from ayahuasca, also known by the name "caapi."

Another research student of Schultes was **Mark Plotkin** (born in 1955), well known for his conservation studies on the South American rainforests. Plotkin's 1993 publication, *Tales of a Shaman's Apprentice,* helped raise public awareness about the importance of conserving indigenous knowledge and wisdom. Plotkin has spent about twenty-five years living and working closely with shamans in South America. He established a nonprofit Amazon Conservation Team in collaboration with Liliana Madrigal of Costa Rica in 1995. One of the goals of this team is to conserve the traditional wisdom of indigenous people from being lost in modern society. Plotkin's work is mainly based on Suriname, Colombia and Brazil.

Michael J. Balick (born in 1960s) also did research under the supervision of Schultes. Balick has spent nearly thirty years of his life carrying out ethnobotanical fieldwork based in China, Egypt, India, Israel, Thailand, and many parts of Central and South America. In 1981 he co-founded The New York Botanical Garden's Institute of Economic Botany, the largest in the nation, where he serves as the Vice President

Fig. 1.3b Michael J. Balick
(Dr. George Milne/The New York Botanical Garden)

and Director. He is also the co-founder of Ix Chel Tropical Research Foundation. He has authored fifteen books on the benefits of rainforests and the relationship between indigenous cultures and their environment, including *Rainforest Remedies* and *Plants, People and Culture* published in 1996. From 1986 to 1996 he helped lead a collaboration between The New York Botanical Garden and the National Cancer Institute in the United States to survey Central and South America and the Caribbean for plants with potential applications against cancer and AIDS.

Several other ethnobotanists around the world have contributed to this field and continue to do so.

Training and Methods

The field of ethnobotany requires a variety of skills. To start with, one must obtain basic training in botany and anthropology with a few courses in plant taxonomy. Plant taxonomy provides the taxonomic training to identify and categorize plants and teaches the ethics of field collection and the methods involved in preparing voucher specimens. In addition, the ability to endure fieldwork, a genuine interest in learning the cultural aspects of plants and people, and great social skills should be sufficient for an amateur ethnobotanist. If you want to take up a career in ethnobotany, you will need extensive training in conservation botany and taxonomy, some linguistic training, years of commitment to fieldwork, and the ability to collaborate with experts in other fields.

A prospective ethnobotanist should follow the basic rules in ethnobotany. If you plan to take up ethnobotany for a career, remember to follow the guidelines given below.

- ◆ Locate the area of ethnobotanical interest.
- ◆ Obtain necessary legal permits to visit and travel in the area.
- ◆ Create a trusting relationship with local people by respecting their space and traditions.

◆ Observe the indigenous uses, preparation methods, and preservation techniques.

◆ Interview the tribal group authority or knowledgeable member to obtain more information.

◆ Be wary of people who would make up false stories to get attention.

◆ With permission, collect plants of interest and make scientifically prepared voucher specimens. These are called herbarium specimens.

◆ Do not collect the plants if they seem rare or endangered.

◆ Identify the plants precisely. Remember proper identification is an important aspect of this field.

◆ Your identification must include the scientific name and family at the minimum. If necessary, get help from taxonomists who specialize in the group.

◆ Voucher specimens must include a description of the plant, its distribution, and taxonomic data to help in identification and verification later on.

◆ Take descriptive notes and supplement with photographs and drawings.

◆ Screen the material for bioactive compounds if it is a potential medicinal source.

◆ Follow the original methods of preparation by the indigenous people. They have years of wisdom passed from previous generations.

◆ Test the active compounds and conduct pharmacological evaluation.

◆ Get the patent rights and clinical trials.

◆ If your plant is a potential future useful source for food, drug, or other valuable products, follow the ethics of ethnobotany and include the indigenous people in a contract and profits.

Ethnobotany in Your Life

This textbook focuses on all types of cultures, from the past to the present and from the indigenous peoples of the Mayan civilization to the immigrants of all continents who have incorporated the plants of their own cultures and those of other cultures into their lives, with modifications to suit their own lifestyles and benefit the rest of human population. For example, the bitter spicy sacred beverage of the Aztec tribe has become the delicious chocolate of our time; the papyrus of the Egyptians has led to the discovery of the modern paper industry; the techniques of sugar production have crossed the borders of India; and researchers are transforming corn into plastic and biodiesel.

As you are already aware, food is an important part of your daily life. In order to understand the role of ethnobotany in your diet, imagine having a simple snack—***vegetarian maki sushi***—like the one pictured above. Unfortunately, a plant book allows no meat! Please note that there are at least ten plant products involved in making this simple snack. Add a beverage like green tea, a small oatmeal cookie and a paper towel to clean up after the snack, and you increase the number of plant products. All of this while you are comfortably seated in your own room or in a sushi bar. You don't even have to fly to Japan to taste their indigenous maki sushi or to China to get the soy sauce. Welcome to modern ethnobotany! Not a single ingredient in your favorite snack is indigenous to the United States where you are most likely to be while eating it.

Plants and Health

In recent years a healthy trend is developing among informed people, and college students are no exception. We are looking for food that will keep us free from bad cholesterol; provide us adequate amounts of fiber, protein, vitamins, and minerals; and at the same time satisfy our taste buds. Health conscious people are consuming more and more plant products and reducing the intake of meat. Throughout this textbook, you will encounter brief mentions of the nutritional benefits of certain plant products. In order to help you understand some of the commonly used terminology like "antioxidant" or "soluble fiber," this book also provides the following simple descriptions and examples. You will find more details and examples throughout the book.

Carbohydrates available from plants include fructose, sucrose, maltose, starch, and cellulose. Glucose forms when we digest these carbohydrates. Fruits and corn syrup contain fructose; cane sugar is an example of

| 1. Rice | 3. Celery | 5. Avocado | 7. Ginger | 9. Bamboo mat |
| 2. Sesame seed | 4. Carrot | 6. Wasabi | 8. Soy sauce | 10. Cotton thread |

Fig. 1.4 Plant ingredients in a vegetarian maki sushi set up
Courtesy of Gopi Vajravelu

sucrose; maltose is in grains like barley; and starch is abundant as a storage product in rice and white potato. Most cultures consider a diet high in sugars and starch unhealthy. A structural form of carbohydrate is cellulose; this along with lignin and pectin form the basis of dietary fiber. Most plant **fibers**, such as cellulose and lignin, are **insoluble**; they do not provide any direct nutritional benefits other than adding bulk and roughage to the diet that is essential for a healthy digestive system. **Soluble** fibers found in pectin, mucilage, and gums reduce the amount of low density lipoproteins (LDL) in the blood. Whole grains, fruits, and vegetables provide insoluble fibers; apple and oats are rich in soluble fibers. An adequate intake of soluble fibers helps in the overall health of the digestive tract. It also maintains blood glucose levels and prevents colon cancer.

Proteins are important for overall growth, health, and maintenance of adequate metabolism. Proteins derived from vegetables, nuts, and grains have more diverse amino acid patterns. Legumes like soybeans, lentils, and chick peas provide high quality proteins and essential amino acids. Lysine and tryptophan are essential amino acids found in quinoa and chocolate respectively. Mixing up a variety of grains, legumes and nuts during various meals of a day would ensure a protein-rich diet for vegetarians.

Plants do not contain cholesterol or trans fats by themselves. However, eating a high **fat** diet can lead to high cholesterol levels over time, and trans fats are modified fats that are unhealthy even if derived from plants. Vegetable oils in general refer to all the oils derived from plants. Most vegetable oils, such as olive oil and walnut oil, contain **unsaturated fats** that are also good sources of essential fatty acids that the human body cannot synthesize but must obtain from diet. There are a few exceptions like coconut oil and palm oil that are loaded with **saturated fats**; regular consumption of saturated vegetable fats will contribute to cholesterol build up. Chapter 7 focuses entirely on vegetable oils.

Calcium and iron are **minerals** that we can obtain from plants as well. Calcium is an abundant mineral that is very important for bone health, and we need an average of 1000 mg per day. Artichoke, broccoli, kale, spinach, asparagus, orange, blackberries, and most nuts are rich in calcium. Iron is a trace mineral but still essential for overall health. We need 10–15 mg a day, Iron deficiency will result in anemia and impaired learning ability. Dark green vegetables, dried fruits, sweet potato, legumes, whole grains, enriched breads and cereals provide nonheme iron. While certain factors like oxalates and calcium may interfere with the absorption of nonheme iron, vitamin C and sugars from foods eaten at the same meal may enhance the absorption. Copper is necessary to absorb iron, and we need about 1.5–3.0 mg per day. Kiwi fruits and lima beans are significant sources of copper. Potassium is necessary for the body's growth and for maintaining water balance in the body. We need about 2000 mg per day. Deficiency results in muscular cramps and weakness. Banana, avocado, and most nuts are good sources of potassium. Selenium functions as an antioxidant in conjunction with vitamin E; we need an average of 60 mg per day. Most nuts provide significant amounts of selenium. Selenium deficiency is rare in humans. Zinc is necessary in negligible amounts, about 10–15 mg per day. Deficiency results in stunted growth. Most nuts have some zinc; kiwi and blackberries, lima beans and corn provide a significant amount.

There are a number of other minerals found in plant foods. You can find them listed under many such sources. If you need more information the following resource will be helpful: **USDA Food & Nutrition Center**.

Plants are excellent sources of natural vitamins and minerals. There are two kinds of vitamins: **water soluble vitamins** that include vitamin C, and a variety of B complex vitamins: thiamine, riboflavin, niacin, biotin, folic acid, and so forth. Folic acid is especially important for vegetarians because its deficiency causes anemia in people who do not consume iron containing meat products. Fresh fruits, legumes, whole grains, seeds, and leafy green vegetables are rich sources of water soluble vitamins. **Fat soluble vitamins** are vitamins A, D, E, and K. Except for vitamin D, which our body synthesizes by utilizing sunlight or obtains from animal sources, we can find the other three vitamins in seeds, fruits, and dark green or yellow vegetables, such as carrots, broccoli, and spinach. Finally, here is some basic information on **antioxidants**. These are phytochemicals that protect the body against certain types of life threatening diseases by fighting against free radicals in our blood that might cause stroke, heart disease, Alzheimer's disease and cancer. Antioxidants also help prevent and repair damage caused by the free radicals. Carotenoids and flavonoids are the two classes of antioxidants.

Carotenoids are found in yellow, orange, red, and green vegetables like carrots and broccoli and in oranges, strawberries, whole grains, and nuts in the form of vitamins A, C and E. **Lycopene** is a carotenoid that is abundant in tomatoes. **Lutein** is abundant in spinach, squash, and carrots. Researchers have proven that consuming such plant products is beneficial in reducing heart disease, stroke, and cancer.

Flavonoids (polyphenols) are often found abundantly in fruits and vegetables that are red, blue, purple, or black like blueberries and grapes. Some other good sources are soybean, pomegranate, tea, and cranberries. **Quercetin**, found in red onions and apples, is a powerful natural antioxidant. Similarly, **resveratrol**, found in purple grapes and **catechins** found in green tea are known for their antioxidant benefits. Flavonoids are beneficial in improving overall health, especially with aging and memory, as well as in reducing the risk of cancer and heart disease.

Plants in History

As we mentioned earlier, people started to bring plants to new territories as the need for them increased. When the plants could not thrive well in new places, the next alternative was to start a trade for personal and financial gain. If the trade was competitive, the next move was to capture and own the land. Such incidents have frequently occurred in history, not necessarily in that order. It is interesting to note how plants have played a key role in creating major wars that changed the fates of countries and the lives of many people. It is even more interesting to note that many plants introduced from other places caused such impacts.

One such plant was **tea**, which originated in the orient at the foothills of Himalayas. It was the symbol of sophistication and social elegance in the orient. When tea arrived in England and then later in the United States during the seventeenth century, the duty imposed on British Tea aggravated the colonial merchants and led to an important historical event called the "Boston Tea Party." This united the colonists and eventually led to the American Revolutionary war and the Declaration of Independence.

Sugarcane was originally distributed in the islands of the Pacific Ocean and adjacent tropical nations until Alexander the Great eyed it around 350 BCE. He brought back stories of how Indians made honey out of reeds. Imported sugar was not enough to satisfy the sweet teeth of Western populations; they had to grow sugarcanes locally. So it was planted across the Mediterranean and was introduced to the New World through Columbus. The successful establishment of sugarcane fields needed manpower that people were bought by exchanging goods and products made out of sugarcane. By the 1700s approximately 16,000 Africans had become enslaved and transported to work in the sugarcane fields. The (in)famous American sugar triangle was one of the main reasons for the enslavement of Africans in Brazil, Cuba, and Haiti.

Opium changed China's history in the 1800s. When China refused the opium trade from Britain in exchange for their silver and tea, the Opium War erupted; it ended with Hong Kong becoming a British colony after China's defeat. For nearly three hundred years, opium ruled China's fate in history, until China shut its borders on all outside trade after it became The People's Republic of China in 1949.

Cotton changed the history and the economy of India and Britain between the eighteenth and twentieth centuries. Cotton from India was a much sought-after commodity until the establishment of the British East India Company. During the colonial rule established by the British in India, cotton processing and manufacturing declined. This was one of the main reasons for the non-violent freedom movement led by Mahatma Gandhi that eventually obtained India's independence from the British. Many of our everyday terms related to cotton clothing, such as khaki, madras, calico, chintz, sash, and shawl, derive originally from Indian languages. The New World species of cotton played a significant role in the history of the United States. An alternate for Indian cotton came from the West Indies and Brazil. Southern parts of the United States started to establish cotton plantations, to employ slaves, and to invent machinery to spin and weave cotton locally. A common saying is that America was built on cotton and was nearly destroyed by it. Senator Henry Hammond in 1858 coined the phrase "cotton is king," referring to the dependence of the world economy upon cotton plantations in the American South.

The **White potato** that originally came from South America impacted the life of the Irish in the middle of the nineteenth century. The Great Irish famine drove people out of Ireland, many of whom settled in the United States.

The historical explorations of Marco Polo, Vasco da Gama, and Magellan identified the southeastern countries and islands for the Portuguese, Dutch, Spanish, and British to conquer and rule. Wars over the **Spice Islands** broke out in the fifteenth century and continued for about two hundred years. Finally, as we all know, an entrepreneur who set out on a voyage in search of Indian spices accidentally discovered the New World along with North America.

In recent years, introduced plants have played significant roles in the economies of other nations. Farmers grow the Chinese **soybean**, for example, extensively in the United States while, on the other hand, the South American **peanut** is ruling in China. Introduced **bananas** increased the financial status of the Caribbean countries. The list would go on, but the message is clear, plants were important to ancient people and remain the same in modern society.

Economically Important Plant Families

Researchers estimate that there are about 350,000 plant species on earth out of which about 260,000 are plants that produce flowers, fruits, and seeds, called flowering plants. Historically, humans have utilized about 2,000 plant species in their lives. Out of these, people have extensively cultivated and commonly utilized only about 10 percent in the entire world. Botanists have traditionally grouped plants into families. As you can see below in Table 1.1, a novice student can identify family names by the ending "ceae." There are taxonomic reasons for

TABLE 1.1 Economically important plant families

Family Name	Common Name	Ethnobotanical Uses	Some Included Members
Amaranthaceae	Amaranth family	vegetables, quinoa, sugar	spinach, sugar beet
Apiaceae	Carrot family	vegetables, poison, herbs	carrot, cumin
Arecaceae	Palm family	fruits, oil, fiber, cosmetics	date, coconut
Asteraceae	Sunflower family	vegetables, oil, dye, medicine	lettuce, sunflower
Brassicaceae	Mustard family	vegetables, herbs, oils	mustard, broccoli
Fabaceae	Bean family	vegetables, oil, sauce, plastic, preservatives	beans, peanut
Musaceae	Banana family	fruits, fiber, religious rituals	banana
Poaceae	Grass family	grains, paper, wood, cooking oil, sugar, biodiesel, plastic	wheat, corn
Rosaceae	Rose family	fruits, nuts, ornamentals	apple, almond
Rubiaceae	Madder family	beverage, medicine, dye, fruits	coffee, cinchona
Solanaceae	Nightshade family	vegetables, medicine, drugs	tomato, tobacco

the way the plants and their families are named that Chapter 2 briefly addresses. There are about 120 families that include plants used for food, medicine, cosmetics, textiles, construction, paper, and spices, in religious rituals, and more. Out of these the following families listed in Table 1.1 stand out from the rest by giving us a variety of useful products.

Most widely cultivated crops

The choice of most widely consumed food crops varies between countries and their inhabitants. Based on local demand, people either cultivate food crops locally or import them from other countries. Out of 150 plants widely cultivated around the world, the following ten crops stand out as those most consumed. The map indicates the original places of origin. Numbers occur only for reference, not in order of importance or as the precise location.

TABLE 1.2 A sample of most widely cultivated crops

Common Name	Binomial	Native origin
1. Rice	*Oryza sativa*	Southeast Asia
2. Wheat	*Triticum monococcum*	Iraq, Syria
3. Corn	*Zea mays*	Mexico
4. Sugarcane	*Saccharum officinarum*	New Guinea and Pacific Islands
5. Banana	*Musa* spp.	Australia
6. Cassava	*Manihot esculenta*	South America
7. Sugar beets	*Beta vulgaris*	Mediterranean countries
8. Soy bean	*Glycine max*	Northeastern China
9. White potato	*Solanum tuberosum*	South America
10. Sweet potato	*Ipomoea batatas*	South America

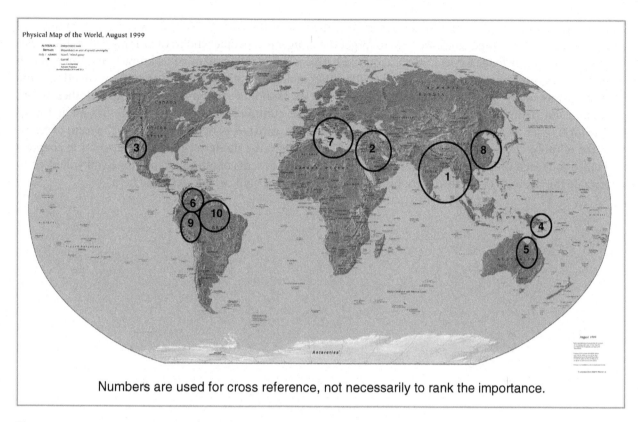

Numbers are used for cross reference, not necessarily to rank the importance.

Fig. 1.5 A sample of most widely cultivated crops and their places of origin

Just for Fun

Some useful sites for students to explore for fun:

Centre for International Ethnomedical Education and Research
www.cieer.org/

Guide to Economic Botany Links
www.rbgkew.org.uk/scihort/eblinks/

Project Open Hand
http://www.openhand.org

Tropicos
www.tropicos.org

Something to think about

As people of a modern generation, we are buying our food supplies from air conditioned supermarkets and medicines from well-organized pharmacies; and our readymade clothes are already waiting for us in fancy stores. In our busy lifestyle, our biggest concern is to get the best products money can buy and to leave the store as quickly as possible only to get busier with the rest of the day. As we put away healthy grains and pre-measured doses of medicines, we tend to forget the indigenous people who in their remote tribal societies have saved this knowledge for us in their acquired native wisdom. How many of us know that quinoa, much valued as a high-protein grain today, and the 50 million tablets of aspirin Americans swallow in a day, came through ethnobotany? Besides food, other products that we cannot live without are beverages, clothes, paper, medicine, and cosmetics that had their origins in different cultures of the world. Society has made things simple for us over the years, even though we tend to complain otherwise. In most cases it is the team of scientists, the ethnobotanists, who brought such tribal knowledge to the modern public. The upcoming chapters will talk about each of these plants and products in detail. As students of ethnobotany, hope you will learn to appreciate the dedication of those ethnobotanists who have shed light on this dynamic field.

Review Questions

Give brief answers in the given space.

1. Define and differentiate between ethnobotany and economic botany.

2. Outline the contributions of Richard Schultes to ethnobotany.

3. Name at least three plants that made history. State their origins and the impact they caused.

4. Outline the basic skills needed for an amateur ethnobotanist.

5. Why is it necessary to have plant names properly identified?

6. What is the reason behind making voucher specimens in the field?

7. What are soluble and insoluble fibers in diet?

8. Explain the benefits of antioxidants in human diet.

9. Which plant family gives us a variety of products ranging from food to biodiesel? Name some members of this family.

10. Reflect on a single plant product that you used today that belongs to one of the listed families in Table 1.1. Write the product and the family name.

Multiple Choice Questions

For each question, choose one best answer.

1. Which one of the following ethnobotanists wrote the book Tales of a Shaman's Apprentice?
    A. Richard Schultes
    B. Michael Balick
    C. Edmund Davis
    D. Mark Plotkin
    E. Robert Raffauf

2. Which plant crop was the main reason for "The Great Irish Famine"?
    A. Sweet potato
    B. Sugar cane
    C. White potato
    D. Tea
    E. Both A and C

Match the following plants from Column 1 to the places of their origin given under Column 2. You may use a choice once, more than once, or not at all.

Column 1	Column 2
3. White potato	A. Pacific islands
	B. South America
4. Sugar cane	C. Himalayan foothills
	D. Africa
5. Peanut	E. Ireland

6. The storage form of carbohydrate in plants is:
    A. fructose
    B. cellulose
    C. starch
    D. fiber
    E. both B and C

7. Which vitamin do our bodies synthesize by using sunlight?
    A. Vitamin A
    B. Vitamin C
    C. B complex vitamins
    D. Vitamin D
    E. Vitamin E

8. Lycopene is _____.
    A. a flavonoid
    B. an antioxidant
    C. a carotenoid
    D. all of the above
    E. both B and C

9. What are the necessary skills of an ethnobotanist?
    A. Plant taxonomy
    B. A basic knowledge of anthropology
    C. Endurance in the field
    D. Social skills
    E. All of the above

10. Which plant family includes beans and peanuts?
    A. Rosaceae
    B. Amaranthaceae
    C. Asteraceae
    D. Fabaceae
    E. Poaceae

True/false Questions

1. Economic botany focuses only on the cultural aspects of plants.   **True**   **False**

2. Richard Schultes' research was on the hallucinogenic plants of South Africa.   **True**   **False**

3. The major responsibility of an ethnobotanist is only to collect the plants.   **True**   **False**

4. Rainforest Remedies is one of the publications of Michael Balick.   **True**   **False**

5. When undertaking ethnobotanical collections, one should collect all the plants, including the rare species.   **True**   **False**

6. Shawl is an American Indian name for cotton.   **True**   **False**

7. Insoluble fibers come packed with huge nutritional benefits in human diet.   **True**   **False**

8. Vitamin C is a water soluble vitamin.   **True**   **False**

9. Dark purple or black colored fruits usually come packed with flavonoids.   **True**   **False**

10. Apples and almonds both have a botanical relationship with roses, and are included in the same family.   **True**   **False**

CHAPTER 2
BOTANY IN A NUTSHELL

Botany is the study of plants and plant life. The word derives from a Greek word, *Botane*, which means plant or herb. Plants are a major group of organisms capable of transforming atmospheric carbon into usable forms of energy by utilizing light and chlorophyll. Plants sustain life on earth by this process, which we call *photosynthesis*. Through the virtue of this chemical process, plants can take care of their own energy needs and provide for the rest of the organisms.

There are four major groups of plants: **Angiosperms** (flower-bearing seed plants); **Gymnosperms** (cone-bearing seed plants); **Seedless Vascular Plants**, which include club mosses and ferns; and the most primitive **Bryophytes**, which include peat mosses and sphagnum mosses. Scientists estimate that there are 350,000 species of plants in all of the four groups mentioned above, of which about 259,000 belong to angiosperms. Flowering plants are the most successful, abundant, and diversified group, as well as the most recent, on earth. They have several adaptations to live on land and are the most advanced plant group of all. As human beings, we utilize plants in all aspects of our lives every single day. In order to appreciate the value of plants it is important to learn the structure and organization of the plant body.

The majority of the plants discussed in this textbook are angiosperms. Parts of a typical flowering plant are described below. Many of these plant parts end up as your dinner or beverage; some help you feel better as medicine; others provide the fiber you wear as clothing, yet many give you wood and other products.

Angiosperm Plant Structure

Most angiosperm plants grow from a seed. A seed consists of an **embryo** and a source of nutrients, the **cotyledon(s)** and the **endosperm**, all protected by two layers of seed coat called the **integuments**. Some seeds like corn and coconut have a well developed endosperm compared to other seeds, such as beans that exhaust their endosperm at early stages of seed development. As the seed germinates, the **hypocotyl** region of the embryo emerges first and becomes established as the primary root. The **epicotyl** region emerges later and becomes the shoot system. The shoot system is above the ground level and is photosynthetic for the most part, and the root system is below the ground level. There are exceptions to this in some plants that have portions of their shoots or roots located in opposite territory and that become modified for various reasons. Chapter 3 describes some of these modifications.

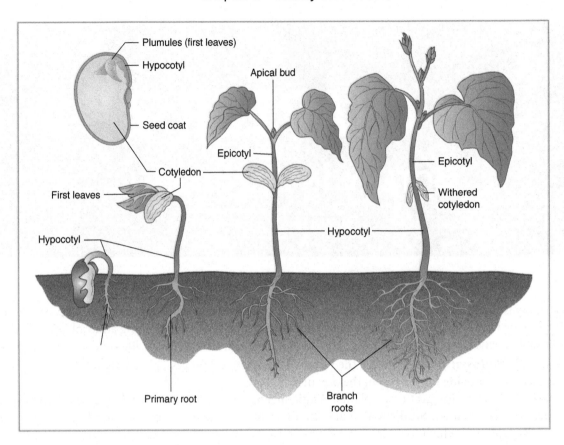

Fig. 2.1 Seed germination
© Kendall/Hunt Publishing Company

Morphology of a Twig

A twig consists of an **axis** (a portion of the stem or a branch) with attached leaves. The **node** is the point on the axis where the leaf attaches. The region on the axis between two such nodes is the **internode**. The length of the internode determines the distance between the leaves. The closer the nodes, the denser the leaves are on the axis. A leaf usually has a flattened, green **leaf blade**. The overall shape of the leaf blade is unique for each species of plant, and the veins that run through the leaf blade largely determine that shape. The leaf blade attaches to the axis either directly (known as **sessile**) or through a stalk called the **petiole**. Some plants, such as the *Hibiscus*, have a pair of short appendages called **stipules** on either side of the leaf petiole. The exact purpose and function of the stipules are not clear, but they may contribute to the photosynthetic capacity of the leaf blade. We call the angle formed between a petiole and the axis an **axil** and the bud located in the axil an **axillary bud**. The new season's branch or flower emerges from this axillary bud. The extreme tip of the growing plant is the **apex**. Botanists call the bud present at the apex an **apical bud**, or a terminal bud. Apical buds contain actively dividing cells known as **apical meristems** that help lengthen the plant during the plant's growing season, or to produce flowers.

Morphology of a Root

Roots anchor the plants to the soil and absorb and conduct water and minerals through specialized tissues. Roots do not show the differentiation of nodes and internodes, nor do they have any buds. The primary root is either well developed with many branch roots that form a **tap root** or it disintegrates and several uniformly

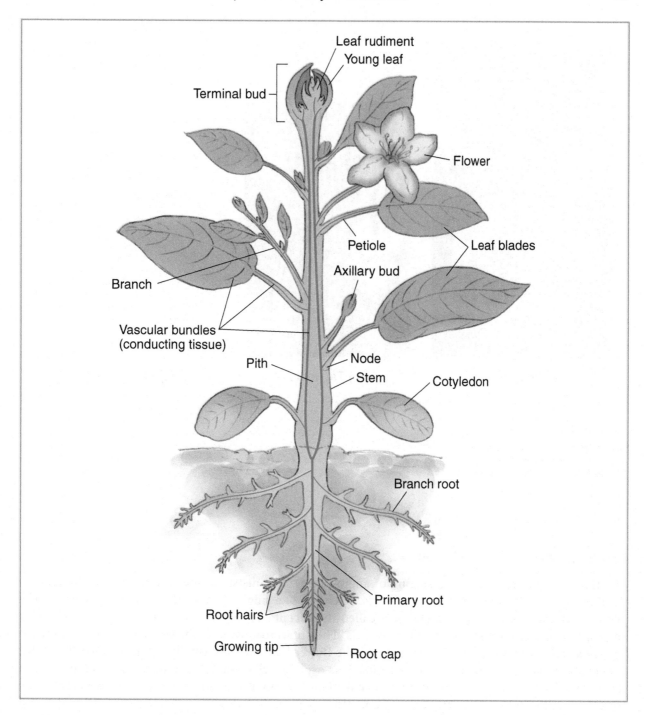

Fig. 2.2 Plant structure
© Kendall/Hunt Publishing Company

thick **fibrous roots** eventually replace it. Fine root hairs found at the root tips help absorb water and minerals from the soil.

Some plants such as corn develop additional roots that emerge from other parts of the plant than the root system. Botanists call such roots **adventitious roots**. There are several types of adventitious roots, including the prop roots of the banyan tree.

(a) Prop roots   (b) Tap roots   (c) Fibrous roots

Fig. 2.3 Types of roots

(a) © Peter Etchells/Shutterstock.com
(b) © Cary Bates/Shutterstock.com
(c) © Pelevina Ksinia/Shutterstock.com

Morphology of a Flower

The flower is the reproductive structure of angiosperms. Botanists consider it a modified shoot where the floral whorls derive from modified leaves. The stalk of a single, solitary flower is the **peduncle**. The peduncle usually expands into a platform called the **receptacle** that holds the floral whorls. A typical angiosperm flower consists of four floral whorls. The outermost whorl is the **calyx**, made up of individual units called the **sepals**. The calyx protects the flower bud and is normally green in color. The calyx is **persistent** and stays along with most fruits. This explains why you notice a whorl of green parts on a strawberry. The second whorl of parts is the **corolla**, made up of individual units called the **petals**. The corolla normally appears brightly colored and attracts the pollinators. Together we know the calyx and corolla as the **perianth** and consider them the **non-essential parts** of a flower, as they have no direct involvement in reproduction. In flowers such as the Easter lily, the perianth is undifferentiated and made up of individual units called the **tepals**. A flower can still go through the processes of pollination and fertilization without the help of the perianth. For example, wind pollinated grass florets (tiny flowers) have undeveloped perianth and still manage to complete the reproduction process. In addition to the perianth, some flowers have additional leaf-like structures called **bracts** found outside of the calyx. Bracts are modified floral leaves that are normally green, but in some flowers, such as in *Bougainvillea,* they also may have attractive coloration.

The third whorl of the flower is the **androecium** (derived from the Greek andro = male/man; oecium = household), made up of individual units called **stamens**. A stamen consists of a stalk called a **filament** that ends in a two- to four-lobed **anther**. The anther is the site of pollen formation. Pollens carry the sperms to the eggs. The final and centermost floral whorl is the **gynoecium** (from the Greek gyno = female/woman; oecium = household), which consists of individual units known as the **carpels**. (Several carpels make up a pistil, a term that botanists use interchangeably.) A carpel consists of a reproductive unit called an **ovary**, where one or more **ovules** form. Ovules develop the eggs. The carpel also has a receptive part known as a **stigma** to collect and hold the pollens and a **style** that connects the stigma and ovary. Since the androecium and gynoecium have direct involvement in reproduction by producing sperms and eggs respectively, botanists consider these two floral whorls **essential parts** of the flower.

© Lotus Images/Shutterstock.com

The floral description above is typical for most plants. However, plant species exhibit a variety of structural alterations that make each species unique. Each of the above mentioned floral whorls consists of one to many individual floral parts that exist either as free or as fused to one another. For example, roses consist of several individual petals whereas *Datura* flowers (Jimson's weed) have five fused petals making up a fused corolla tube and free corolla lobes. Stamens of tulips are free while *Hibiscus* stamens fuse into a separate staminal tube. The variation in floral parts brings diversity and adds to the aesthetic value of the flowers.

Botanists describe a flower that consists of both carpels and stamens as **bisexual** or as a **perfect flower**. When any one of the above two parts is missing in a flower, they then describe it as a **unisexual** or as an **imperfect flower**. A unisexual flower may be a staminate flower when the carpels are either missing or non-functional; likewise, a carpellate flower lacks functional stamens. When a flower contains all four whorls, we consider it a **complete flower**; when a flower lacks any one or more whorls, it is an **incomplete flower**. Thus, oak flowers that have separate male and female

© bochimsang12/Shutterstock.com

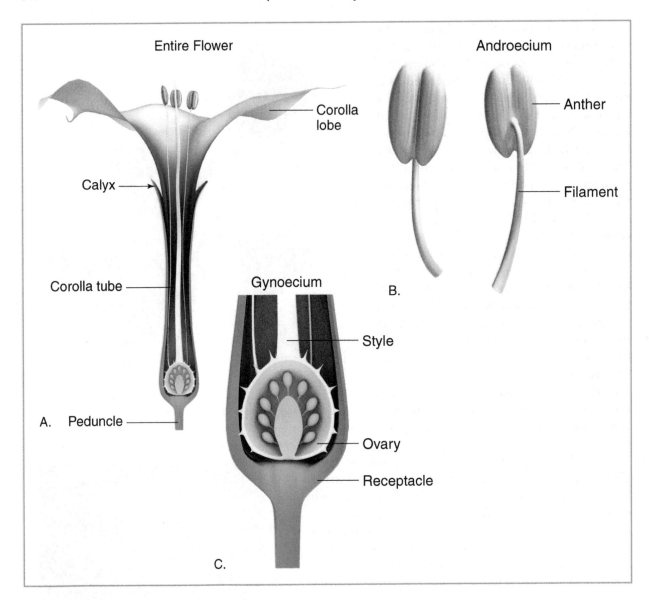

Fig. 2.4 Flower parts of *Datura*
© sciencepics/Shutterstock.com

flowers are not only imperfect, they are also incomplete. Grass florets that have both stamens and carpels but a reduced perianth are perfect and incomplete.

When the flowers are unisexual they may grow on the same plant (separate male and female flowers in one individual plant), a condition called **monoecious** (= one house) as in walnuts; or they may grow on separate plants (male plants and female plants), a condition called **dioecious** (= two houses). If you are growing date palms, make sure you have both male and female plants in close vicinity to ensure pollination and fruit formation because they are dioecious.

So far, we have looked at the details of a single flower structure. Most often flowers are found in collections called an **inflorescence**. The peduncle, instead of terminating in a single flower, usually branches

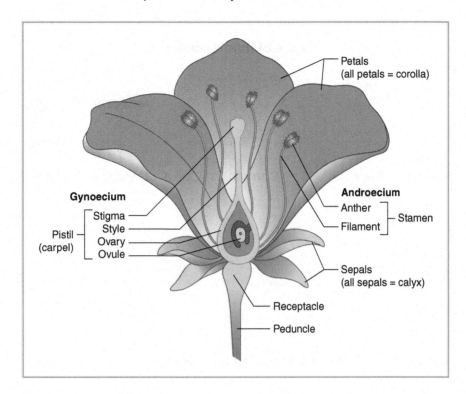

Fig. 2.5 A typical flower in cut view
© Kendall/Hunt Publishing Company

off and holds a collection of flowers arranged in various fashions. In an inflorescence, the stalk of an individual flower is a **pedicel**. The peduncle is the stalk of an inflorescence. As we learned above, the pedicel ends in a receptacle and holds the four floral whorls. There are many different kinds of inflorescences as listed in Fig. 2.6. Thus, a sunflower is a type of inflorescence, known as a capitulum or a head, and not a single flower.

Later in this book we will study different kinds of fruits. After fertilization, the ovary becomes the fruit and the ovules become the seeds. Variations in the fruit are partly due to the position and structure of the ovary. We will see some terminology related to the location of the ovary on the receptacle in relation to the rest of the floral parts, such as the sepals, petals, and stamens.

We may not pay close attention to the location of the ovary within a flower when we look at one, but ovary position is an important factor in the appearance of a fruit. Within a flower, an ovary may rest at a higher level on the receptacle compared to other parts of the flower. We describe such a flower as **hypogynous** because the ovary is in a **superior** position. Example: tomato. In **epigynous** flowers, the ovary is **inferior** and located below the sepals, petals and stamens. Example: blueberry. There are some **perigynous flowers** where the ovary is neither superior nor inferior but somewhat **semi-inferior** (or half-inferior), such as the apple blossom. The receptacle wraps around such an ovary halfway and holds the rest of the floral parts. Whenever you notice a crown on top of fruits like pomegranate, remember that it is the persistent calyx on top of the inferior ovary that has now become the fruit.

Many fruits derive from a **compound ovary** that develops from the fusion of two to several individual carpels. Fruits such as oranges and okra derive from such an ovary. Some fruits derive from **simple ovaries** such as an individual carpel without any fusion. For example, the single compartment inside a bean pod suggests it develops from a single carpel.

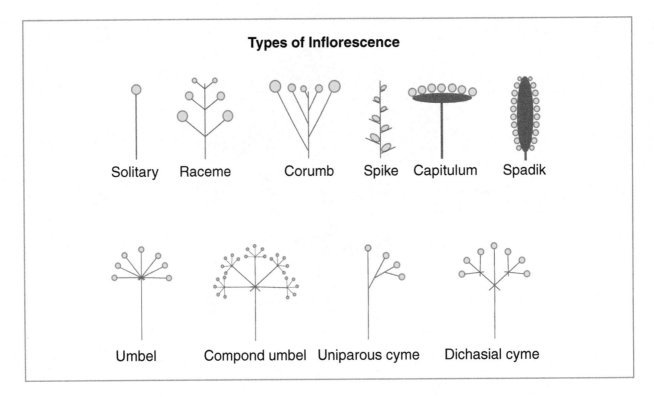

Fig. 2.6 Inflorescence types: Diagrammatic representation
© Nikitina Olga/Shutterstock.com

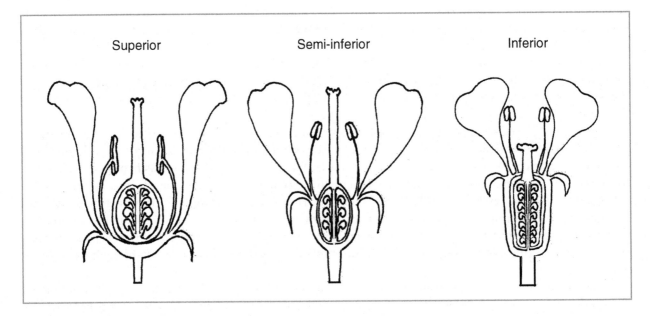

Fig. 2.7 Ovary position: Diagrammatic representation
Courtesy Rani Vajravelu

(a) Bean pods

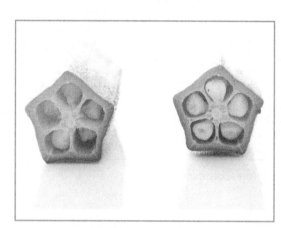

(c) Okra in cut view

(b) Fruit of an okra

Fig. 2.8 Bean and okra

(a) © Lepas/Shutterstock.com
(b) © Philip Yb Studio/Shutterstock.com
(c) © KPG_Payless/Shutterstock.com

Two Classes of Angiosperms

Recent molecular evidence suggests the classification of angiosperms into three classes: monocots, eudicots, and magnoliids. The exact relationship among the three groups is not satisfactorily explained so far. We will continue to use the traditional grouping of two major classes, monocotyledonae (monocots) and dicotledonae (dicots) in this book. Dicots as a group are about twice as large as monocots and include roses, oranges, and pecans. Grasses, palms, and orchids are some of the monocots. The number of **cotyledons** (seed leaves) provides the basis for this classification: one cotyledon in monocots, two in dicots. There are other obvious morphological differences that we summarize in Table 2.1.

TABLE 2.1 Comparison of two classes of angiosperms

Class	Dicotyledonae	Monocotyledonae
Root system	Tap root system	Fibrous root system
Leaf venation	Netted venation	Parallel venation
Flower parts	In fours or fives or multiples of 4 or 5	In threes or multiples of 3
Stem vascular bundels	Arranged in a concentric ring	Scattered throughout the ground tissue
Pollen	3 openings	1 opening
Secondary growth	Present in woody plants	Absent or by primary thickening

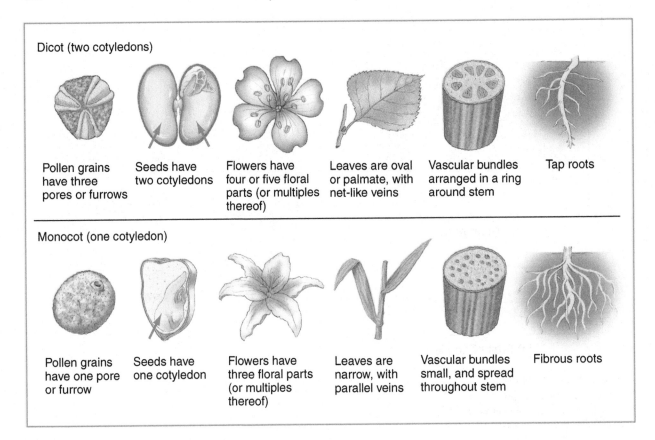

Fig. 2.9 Dicot and Monocot
© Kendall/Hunt Publishing Company

Meristems and Plant Growth

Plants continue to grow throughout their lifetime. Growth in plants occurs by means of actively dividing cells called the **meristems**. Basically, there are three kinds of meristems, and each contributes to a certain type of growth in plants.

1. **Apical meristem:** The location of the apical meristem is at the growing tips of plants, which helps lengthen them. We call such a growth primary growth. Primary growth accumulates primary tissues differentiated from three primary meristems of the apical meristem: **protoderm** (becomes epidermis), **procambium** (becomes vascular tissue), and **ground meristem** (becomes ground tissue). Herbaceous plants and tender growing parts of woody plants undergo primary growth.

2. **Lateral meristem:** This meristem appears as a cylinder running laterally through the entire plant except at the tips, which contributes to the increase in the plant's diameter. We call such a growth secondary growth. **Vascular cambium** and **cork cambium** are two types of lateral meristem. When woody plants need additional tissues, these two meristems divide and accumulate secondary tissues, such as secondary xylem (wood), secondary phloem (part of bark), cork, and periderm (that replaces the epidermis). Raised openings called lenticels also replace stomata.

3. **Intercalary meristem:** The location of this meristem is at the nodal regions of grasses, which contributes to the re-growth of the grasses after cutting.

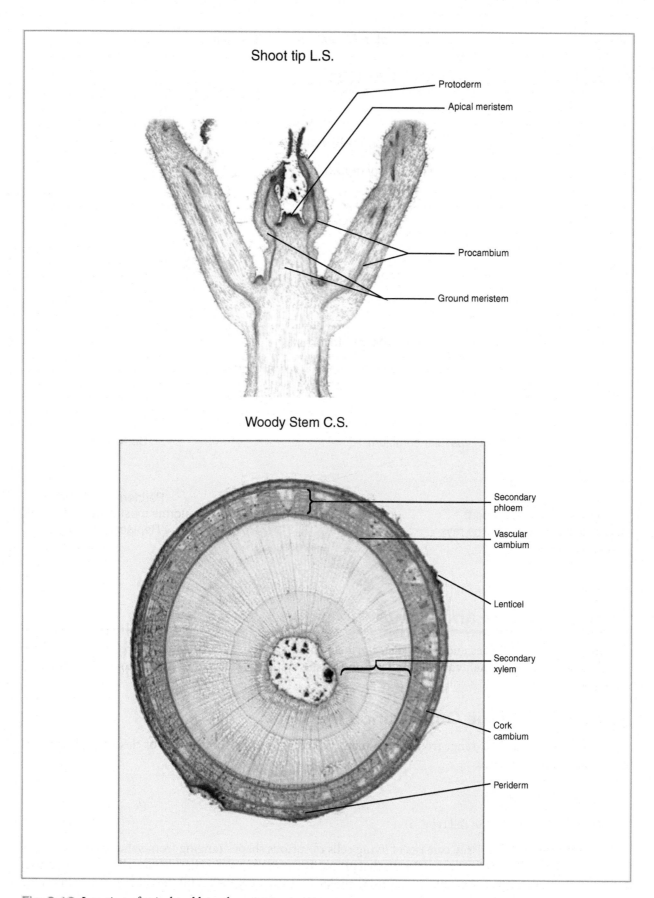

Shoot tip L.S.

Protoderm

Apical meristem

Procambium

Ground meristem

Woody Stem C.S.

Secondary phloem

Vascular cambium

Lenticel

Secondary xylem

Cork cambium

Periderm

Fig. 2.10 Location of apical and lateral meristems in stem

© University of Wisconsin Plant Teaching Collection; Micrographs; © Michael Clayton

Meristems and Derived Tissues

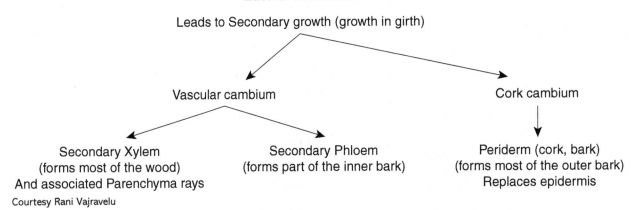

Apical Meristem

Leads to Primary growth (growth in length)

↓

Primary Meristems

↓

Primary Tissues

Lateral Meristem

Leads to Secondary growth (growth in girth)

Vascular cambium

Cork cambium

Secondary Xylem
(forms most of the wood)
And associated Parenchyma rays

Secondary Phloem
(forms part of the inner bark)

Periderm (cork, bark)
(forms most of the outer bark)
Replaces epidermis

Courtesy Rani Vajravelu

Tissue Systems and Included Tissue Types

The apical meristem divides and accumulates primary tissues that in turn become differentiated into three main tissue systems: **dermal**, **vascular** and **ground tissue systems**. Each tissue system consists of different cell types.

Summary of Tissue Organization in Plants

The tissues listed Table 2.2 range from one to many cell components. Another way of classifying them is as simple and complex tissues.

Simple Tissues

Simple tissues consist of one cell type.

> **Parenchyma:** This tissue consists of living cells of various shapes ranging from spherical to fourteen-sided. Cells have primary cell walls that primarily consist of polysaccharide material called **cellulose**. Protoplasm that contains cytoplasm, vacuoles, and other organelles fill the cells. Parenchyma is the most abundant tissue found in plants. Parenchyma cells of leaves and some stems have chloroplasts

TABLE 2.2 Summary of tissue organization in plants

Tissue Systems	Tissues	Location & Function	Component Cell Types
Dermal	Epidermis	Outermost protective layer	Ground cells, guard cells, trichomes
Vascular	Xylem	Part of the vacular tissue; conducts water and minerals	Xylem vessels Xylem tracheids, Xylem fibers, Xylem parenchyma
	Phloem	Part of the vascular tissue; conducts dissolved sugars	Sieve tube elements, companion cells, Phloem fibers, Phloem parenchyma
Ground	Ground tissue Cortex	Region between vascular tissue and epidermis	Collenchyma, parenchyma, sclerenchyma
	Ground tissue Pith	Innermost region of the stem and some roots	Parenchyma

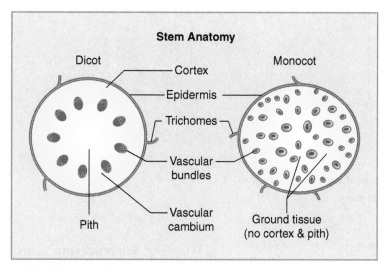

Fig. 2.11 Stem anatomy: Diagrammatic representation
Courtesy Rani Vajravelu

to help in photosynthesis (**chlorenchyma**). Those of aquatic plants have huge air cavities to help them float (**aerenchyma**); and those of roots, fruits, and stems help in the storage of food material, such as starch. When you are eating apples and potatoes you are mostly consuming parenchyma and little of other tissues.

Collenchyma: This consists of living cells like parenchyma except that suberin thickens their primary cell walls, which gives elasticity to the tissue. The location of this tissue is normally in parts of the plant requiring a flexible mechanical support. This tissue is not as abundant as parenchyma. Leaf petioles and tender parts of plant stems contain collenchyma.

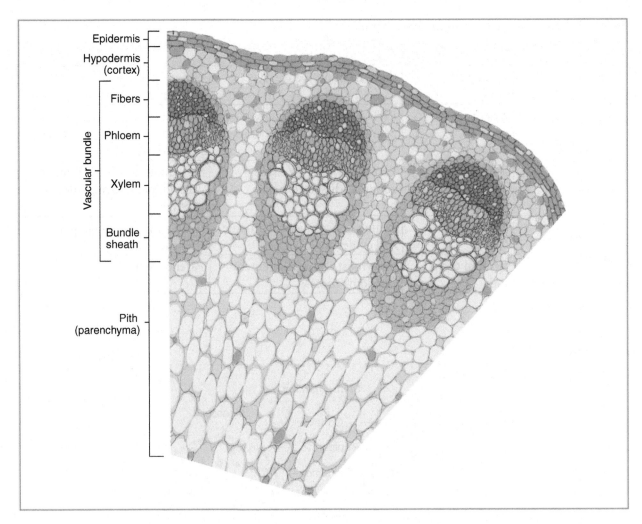

Fig. 2.12 A portion of dicot stem in cross section
© Kendall/Hunt Publishing Company

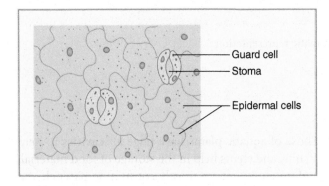

Fig. 2.13 A portion of leaf epidermis
Courtesy Rani Vajravelu

Sclerenchyma: This consists of cells that do not have living protoplasm. At maturity, the cells of this tissue develop thick secondary walls embedded with lignin. This tissue occurs in plant parts that need rigid, mechanical support to compensate for the lack of a skeletal system in plants. Sclerenchyma cells occur in two forms:

Sclereids appear as clusters of stone cells distributed among parenchyma cells of fruits like pear, in the pits of peaches and cherries or as individual branched structures in aquatic leaves.

Fibers appear as elongated, thick walled cells intermingled with other tissues of the plant. Fibers have enormous commercial value including the paper that you are using.

Complex Tissues

Complex tissues consist of more than one cell type.

Epidermis: The epidermis forms the outermost layer of plant organs. It normally consists of a single layer of living parenchyma cells called epidermal or **ground cells** that fit together like pieces of a puzzle. Intermingled with these epidermal cells are a pair of bean shaped **guard cells** that enclose an opening called a **stoma** (pl: stomata). Stomata help in gas exchange and regulate the water evaporation through the leaf surface. Ground cells often have some outgrowths called **trichomes** that protect the epidermal layer along with a waxy layer of **cuticle** that cover the outer surface.

Vascular tissues: Xylem and phloem are major types of vascular tissues in most plants. They conduct water, minerals, and dissolved sugars throughout the plant body. Within stems, xylem and phloem occur close together in the same radius and make up discrete vascular bundles. In roots they occur in separate radii alternating with each other within a vascular cylinder or **stele**.

Xylem: As part of the vascular tissue system, xylem conducts water and minerals. It consists of several specialized cell types including xylem vessels, tracheids, fibers, and parenchyma cells. Botanists generally consider this tissue as dead tissue because the majority of the cells have lignified walls and lack protoplasm. When a plant undergoes secondary growth, a **secondary xylem** accumulates and makes up what we generally call **wood** in a tree.

Phloem: Phloem occurs alongside the xylem and conducts dissolved food substances. Considered a living tissue, it consists of living sieve-tube elements, companion cells, and parenchyma, along with some phloem fibers. After secondary growth, the newly formed phloem becomes **secondary phloem**. The previous year's phloem turns into phloem fibers and become part of the bark.

Secretory Cells

Certain cells of the apical meristem specialize as secretory structures that intermingle with the rest of the primary and secondary tissues in plants. Such specialized cells are responsible for oil, latex, resin, nectar, perfumes, and mucilage found in plants. Milky sap of papaya is a good example of secretory structures found in plant parts.

Plant Life Cycle

Plants go through an entire life cycle starting from a seed that then grows into a stem, leaves, flowers, and fruits. The cycle ends with a new set of seeds. Genetics and environment play an important role in controlling this life cycle pattern for most naturally growing plants. There are three major categories of plants, based on the duration of their life cycles. **Annuals** like most weeds and beans complete their life cycle in one growing season. **Biennials** like carrots need two years. The first year is utilized to complete their vegetative growth and to store food in the underground part that will regenerate during the second year to complete the reproductive part of the life cycle by flowering, producing fruits and seeds. **Perennials** normally take several years to complete their vegetative growth. Once they start their reproductive stage, these plants flower and produce fruits at a certain season every year. All of our fruit trees such as apples and avocados are perennials.

Botanical Nomenclature

Plants are known by local names called vernacular names that may differ between the geographic regions and dialects. For example, ladies finger, gumbo, and vendi are some of the local names used in different parts of the world for a vegetable plant that you call okra. For scientific purposes this plant and all other plants must have one accepted scientific name. Such scientific names avoid ambiguity in plant research and communication. These names derive mostly from Latin or Greek and have two parts: a **genus** (pl. genera) name, followed by a **species** (pl. species) name. Because of this, botanists refer them as **binomials**. Genus is a larger category that might include one to several species. To illustrate this, we will use almond, cherry, rose, blackberry and strawberry as our examples. One genus called *Prunus* includes almond, known by the binomial *Prunus dulcis,* and cherry, *Prunus avium.* This indicates that almonds and cherries are distinct species of their own included within the same genus. Other genera like *Fragaria* (strawberry), *Rubus* (blackberry), and *Rosa* (rose) have their own number of species. All of the above mentioned genera are included in a larger category called **family**. In our example, it is a family called *Rosaceae* (the rose family). Plant taxonomists are the specialists who classify and name the plants based on the conclusions derived from various sources. When referring to any species of a genus in general, it is customary to use the abbreviation sp. (pl. spp.) following the genus names such as, *Prunus* sp., or *Rubus* spp. Even though the scientific names are hard to pronounce and to remember, each name derives from a person's name, place, use, location, shape or color, and so forth.

We must follow the rules and regulations of the International Code of Botanical Nomenclature (ICBN) when we use scientific names. That is why you will notice that we print binomials in italics, that genus names start in capitals, that species names start with a lower case, and that family names end in "ceae." The binomial system came into use during the time of **Carolus Linnaeus**, a Swedish botanist who wrote and used this system in his book **Species Plantarum**, published in 1753. For this reason, we know him as the **Father of Binomial Nomenclature**. Prior to the standardization of this system, one would learn the polynomial *Mentha floribus capitatus, foliis lanceolatis serratis subpetiolatis* for the name of the mint plant, instead of the simple binomial *Mentha piperita* that is in use today.

In the upcoming chapters, we will refer to the plants as vegetables, fruits, beverages, spices, herbs, oils, medicines, wood, fiber, and also as narcotics. You will find these are, in fact, the products of stems, roots, leaves, buds, and tissues of the plants that we learned about in this chapter. You will find a connection between the information given above and your use of plants in everyday life.

Review Questions

Give brief answers in the given space.

1. What are the main differences between the major groups of plants?

2. List all the structures that you would find inside an angiosperm seed.

3. What are the meristems? Briefly explain their function in flowering plants.

4. Describe the structure of a complete flower.

5. List the differences between dicots and monocots.

6. What are the component cell types of a ground tissue system?

7. How would you distinguish a plant that has undergone secondary growth?

8. What is the role of secretory structures found in plants?

9. Why are carrots biennials?

10. Explain the binomial system of nomenclature.

Multiple Choice Questions

1. Which of the following groups is seedless but has vascular tissues?
    A. Gymnosperms
    B. Tracheophytes
    C. Bryophytes
    D. Angiosperms
    E. Both B and C

2. Axillary buds are:
    A. found at the growing tips of stems and roots.
    B. found at the angle between leaf and stem.
    C. the future branches or flowers.
    D. all of the above.
    E. both B and C.

3. If a plant is growing in length, such a growth
    A. is known as secondary growth.
    B. results in wood and bark.
    C. is known as primary growth.
    D. is both A and B.
    E. is both B and C.

4. Vascular tissue generally consists of:
    A. xylem only.
    B. phloem only.
    C. apical meristem.
    D. all of the above.
    E. both A and B.

5. Which statement is NOT generally true of monocots?
    A. They do not undergo secondary growth.
    B. Leaves have parallel venation.
    C. Flower parts are in threes or in multiples of three.
    D. Monocot pollens have a single pore.
    E. The arrangement of vascular bundles is in a concentric ring.

6. The stalk that supports a leaf is the _____, and the stalk that holds a solitary flower is the _____.

    A. peduncle; petiole
    B. petiole; peduncle
    C. node; axillary bud
    D. internode; axillary bud
    E. node; petiole

7. A "perfect" flower

    A. has both non-essential whorls irrespective of the presence or absence of essential whorls.
    B. lacks stamens or pistils, not both.
    C. has both the essential whorls, but always lacks the non-essential whorls.
    D. is also referred to as unisexual.
    E. has both essential whorls irrespective of the presence or absence of non-essential whorls.

8. A staminate flower is also

    A. a male flower.
    B. a unisexual flower.
    C. an imperfect flower.
    D. both A and C.
    E. A, B, and C.

9. Which of the following is the most abundant simple tissue?

    A. Parenchyma
    B. Collenchyma
    C. Sclereids
    D. Sclereids and xylem
    E. Sclerenchyma and phloem

10. Flexible support needed in petiole and tender plant parts is given by

    A. collenchyma cells.
    B. parenchyma cells.
    C. water-conducting cells.
    D. sclerenchyma cells.
    E. food-conducting cells.

True/false Questions

1. A compound ovary derives from two or more fused carpels.   **True**   **False**

2. A flower with semi inferior ovary is also an epigynous flower.   **True**   **False**

3. Monocot stems have vascular bundles arranged in a concentric ring.   **True**   **False**

4. When leaves lack a stalk, we describe such leaves as sessile.   **True**   **False**

5. Carrots are monocots because they have one main root.   **True**   **False**

6. Root hairs take part in the absorption of water and minerals in a root.   **True**   **False**

7. We call the venation characteristic of monocot leaves parallel venation.   **True**   **False**

8. Secondary growth adds to the girth of trunks and roots.   **True**   **False**

9. After secondary growth, the periderm replaces the epidermis.   **True**   **False**

10. Among genus and species, species is a larger category and may include many related genera.   **True**   **False**

CHAPTER 3
VEGETABLES

The term "vegetable" refers to any edible plant part that people consume either cooked or raw as part of their regular meals. Botanically speaking, vegetables derive from vegetative plant parts such as leaves or portions of stems and roots. In the broad sense, vegetables include any plant part prepared as a savory dish rather than as a dessert. This definition includes beans and tomatoes that derive from reproductive parts of the plant and are technically fruits. Vegetables play an important role in the healthy eating lifestyles of people around the world. The definitions and cooking methods of vegetables vary between cultures and traditions. In this chapter, we will divide vegetables into three main categories: stem and leaf-derived; hypocotyl-derived; and those derived from modified plant parts. Chapter 4 will discuss fruit-derived vegetables.

Stem and Leaf-Derived Vegetables

The stem is the main axis of the plant that supports the branches, leaves, flowers, and fruits. The stems of some plants, however, not only perform their normal functions but also have various modifications. In certain plants, stored food may swell the aerial stem or the stem's growth may become stunted with shortened internodes. Both cabbage and lettuce suppress the apical bud growth, which results in a stunted stem growth with tightly packed leaves. The outer leaves, however, expand normally and cover the inner unexpanded leaves, thus forming a headed vegetable. These vegetables are best when allowed to mature at cooler temperatures. At warm temperatures, the stunted stem can expand and produce a flowering shoot. Such "bolting" results in bitter-tasting vegetables and makes them undesirable. Cabbage and lettuce tend to undergo bolting.

The mustard family, Brassicaceae gives us a variety of vegetables and other products, including herbs and oils. We call vegetables derived from Brassicaceae "cruciferous vegetables" based on an old name for the family "Cruciferae." This family includes about three thousand species mainly spread around in temperate regions of the world. Some also grow in tropical areas. Cabbage and other vegetables like kohlrabi, Brussels sprouts, cauliflower, kale, and broccoli are also known as "cole crops" meaning stem vegetables. All of these belong to the genus **Brassica** that is native to the Mediterranean region. All have taproots, which may become fleshy. In spite of their poor reputation, these vegetables are packed with nutrients and antioxidants that have proven to lower LDL cholesterol and reduce cancer causing free radicals in the body. Table 3.1 lists the binomials and places of origin.

Cabbage is one of the oldest vegetables and was an important food item in the lives of European peasants. People have known about sauerkraut, a pickled dish made from shredded cabbage, since 200 BCE. Kimchee is a spiced preparation known in Asian cultures. Many people use both as substitutes for other vegetables, and they are good sources of vitamin C. The common cabbage includes both red and green kinds with smooth leaves; these are the headed varieties. **Savoy cabbage** has loose, crumpled leaves. **Chinese cabbages** come in several varieties, including the bok choy type where people use the stems, petioles, and leaves based on the variety.

Cauliflower and **broccoli** are terminal parts of the stem. Both are relatively recent selections from a common ancestor, a wild form of kale. Massive expansion and proliferation of stem tips by apical meristems form numerous "curds" in cauliflower. Growers traditionally produced white heads by tying the large leaves around the heads to protect them from exposure to sunlight. When exposed to sunlight, they may turn green, which supermarkets sell under the name "broccoflower." Modern varieties of cauliflower lack the ability to turn green even when exposed to the sun.

The vegetable portion of broccoli consists of an extensively branched, tight inflorescence with numerous unopened, fertile flower buds. When not picked at the right time, these flower buds open into yellow blossoms that make the broccoli unsuitable for sale. **Kohlrabi** has a fleshy rounded stem for storage that bears long, stalked leaves. Cooks remove the leaves and thick skin before using it as a vegetable. **Kale** is a form of cabbage in which the stems are not stunted; thus it produces green leaves used as vegetables. The cultivated forms of kale include the collard greens. **Brussels sprouts** result from stunted axillary buds from the main stem. When we remove the leaves from the main stem, the Brussels sprouts appear as miniature heads attached around the stem.

Celery is the petiole portion of celery leaf that is swollen. The genus *Apium* belongs to Apiaceae that also includes carrots and parsley. A variety of celery produces a root vegetable, celeriac, which is popular in Europe.

Another family that provides stem and leafy vegetables is Amaranthaceae. Many tropical parts of the old world cultivate and consume several species of the genus *Amaranthus*. Parts of India, China, and Africa cook the tender stem and protein rich green leaves as a vegetable. These greens are a rich source of vitamins A, C, B6, folic acid, and calcium and are nutritionally similar to common spinach.

Fig. 3.1b Kohlrabi
© Workmans Photos/Shutterstock.com

Fig. 3.1a Broccoli
© Goran Kuzmanovski/Shutterstock.com

The spinach is included in Amaranthaceae under the genus *Spinacea* that people grow widely in the Mediterranean and commonly sell in the United States. Used in place of lettuce in salads, spinach also gets cooked into a vegetable dish. Popeye the sailor, a cartoon character created by E. C. Segar in 1929, has played an important role in promoting the popularity of spinach among American children. It contains a relatively large amount of folic acid and iron. The iron in spinach is a non-heme iron similar to most vegetables, but the oxalic acid present in spinach inhibits its absorption by the human body. The best way to eat spinach is in cooked form along with iron absorption enhancers like vitamin C rich foods.

Another green leafy vegetable is Malabar spinach, *Basella*. The leaves are thick, mucilaginous, and mild tasting. There are two common species mainly based on the color.

Lettuce, *Lactuca sativa*, is in Asteraceae, one of the largest dicot families. Wild forms contain milky latex in tender parts of the plant. Known since 4,500 BCE, lettuce has remained an important vegetable consumed uncooked. A related genus, *Cichorium*, provides endives and chicory. People use both as headed vegetables. In addition, they use the roots of chicory as a coffee substitute.

Fig. 3.1c Brussel sprouts
© Robert Taylor/Shutterstock.com

Asparagus, *Asparagus officinalis*, is in Asparagaceae, a monocot family. The plant is an herbaceous perennial with an underground rhizome. Native to the western coasts of Europe, its use is in cooking and in medicine as a diuretic. Aerial shoots sprout from the rhizomes, and growers harvest the young tip portion for use as vegetables. Mature shoots expand and extensively branch out with whorls of modified cladophylls. Growers asexually propagate asparagus from the pieces of rhizome. Asparagus is a good source of folic acid and dietary fiber.

Fig. 3.2 (A) Amaranth, (B) Malabar spinach, (C) Asparagus

a. © Matjoe/Shutterstock.com
b. © Ajayptp/Shutterstock.com
c. © Phimchanok13/Shutterstock.com

Hypocotyl Vegetables

Hypocotyl is the intermediate portion between the stem and root. Vegetables derived from the hypocotyl often grow partially buried underground and lack a green color. Botanists generally consider them root vegetables because the major portion of the hypocotyl region belongs to the tap roots. Root vegetables within this definition are carrots, parsnips, beetroots, turnips, rutabagas, and radishes.

Carrots, **Daucus carota**, are biennial plants derived from a wild Queen Anne's lace plant. Original domestications of carrots were known to be purple; the recent orange color is the intensified form of a yellow mutant variety. The pigment beta-carotene is responsible for the orange color in carrots. The digestion process releases vitamin A molecules inside the body. People eat carrots both raw and cooked. Parsnips, **Pastinaca sativa**, are pale yellow with a sweeter taste than carrots, and they are not as popular in United States. You can also eat parsnips raw or cooked in soups and stews.

Beet root, **Beta vulgaris**, is in Amaranthaceae. There are three varieties of beets: Swiss chard, a leafy vegetable; fodder beets, a yellowish-white root used as cattle feed; and the sugar beet used as a sugar source. The vegetable type is one of the cultivars that have deep purple roots due to betacyanin pigments. Beets have the highest sugar content of all vegetables and are low in calories. People normally pickle beets or use them in soup. Sugar beets contain a high concentration of sucrose. The U.S., Europe, and Russia cultivate sugar beets

TABLE 3.1 Vegetables derived from stems, leaves, and hypocotyl

Family	Vegetable Name	Binomial	Vegetable Part	Origin
Amaranthaceae	Amaranth greens	Amaranthus tricolor	Stem and leaves	Asia; worldwide in tropics
Amaranthaceae	Beets	*Beta vulgaris*	Hypocotyl	Mediterranean
Amaranthaceae	Spinach	*Spinacea oleracea*	Leaves	Western Asia
Apiaceae	Carrot	*Daucus carota*	Hypocotyl	Europe
Apiaceae	Celery	*Apium graveolens*	Swollen petiole	Eurasia
Apiaceae	Parsnip	*Pastinaca sativa*	Hypocotyl	Mediterranean
Asparagaceae	Asparagus	*Asparagus officinalis*	Tender, stem tips	Western Europe
Basellaceae	Malabar spinach	*Basella alba* *Basella rubra*	Leaves and tender stem	Tropical Asia
Brassicaceae	Broccoli	*Brassica oleracea var. botrytis*	Unopened flower buds	Europe
Brassicaceae	Brussels sprouts	*Brassica oleracea var. gemmifera*	Axillary buds	Europe
Brassicaceae	Cabbage	*Brassica oleracea var. capitata*	Leaves	Mediterranean
Brassicaceae	Cauliflower	*Brassica oleracea var. cauliflora*	Proliferated stem tips	Europe
Brassicaceae	Chinese cabbage	*Brassica Pe-tsai*	Stem and leaves	Asia
Brassicaceae	Collard	*Brassica oleracea var. acephala*	Leaves	Mediteranean
Brassicaceae	Kohlrabi	*Brassica oleracea var. gongyloides*	Storage stem	Mediterranean
Brassicaceae	Radish	*Raphanus sativus*	Hypocotyl	Western Asia
Brassicaceae	Turnip	*Brassica campestris*	Hypocotyl	Eurasia

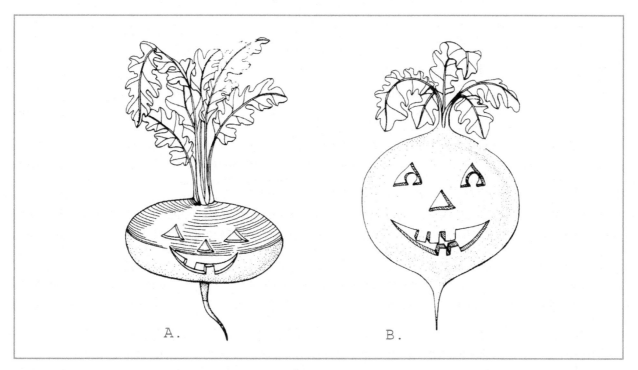

Fig. 3.3 Turnip has a long-time association with Irish traditions. According to an old Irish legend, the first jack-o-lantern was carved out of a turnip. The Irish have used (A) turnips and (B) rutabagas to make lanterns for All Hallow's Day or All Saints' Day. When the Halloween tradition started in America, turnip was replaced by pumpkin.

From *Economic Botany: Plants in our World 3E* by Beryl Simpson & Molly Conner Ogorzaly, © 2001. Reproduced by permission of The McGraw-Hill Companies.

for commercial sugar production, which accounts for 30% of the world's table sugar. Cleaned sugar beets get crushed to remove the syrup. They then undergo distillation and clarification very similar to sugarcane processing. Chapter 5 describes this process under sugarcane.

Turnips, **Brassica campestris**, are one of the oldest crops known, and people have eaten them for about four thousand years. The flesh is white to yellowish with a purple tinge on the skin. Europeans have consumed them through the Middle Ages, and the Chinese in northern China prefer them, but many people, especially in the United States, do not favor them because of their distinctive flavor. Turnips are high in dietary fiber, vitamins C, B6, folic acid, calcium, potassium, and copper. People sometimes use the leaves as turnip greens. Rutabagas, **Brassica napus**, often get mistaken for turnips as they are similar, but they have a stronger flavor and pointed ends. Due to their low moisture content shippers coat their roots with wax while shipping. Cultivars of these two species are used for canola oil.

Radish, **Raphanus sativus**, is another vegetable from the mustard family. There are several varieties of this plant, including the red-skinned version. The daikon or Japanese radish is the winter radish that looks like a white carrot. Radishes are a rich source of vitamin C, B6, folic acid, and fiber. They are low in calories and known for their anti-inflammatory effects. One can eat these root vegetables raw in salads or cooked, which yields a distinctive flavor. People also consume the leaves as vegetable greens.

Fig. 3.4 Hypocotyl vegetables
© Frances L Fruit/Shutterstock.com

Vegetables Derived from Modified Plant Parts

Description of modifications

Stems and roots help in asexual reproduction besides performing their main function of support and conduction. They carry out vegetative reproduction mainly through stored food materials in modified stems or roots underground. A brief description of such modifications follows.

A **rhizome** is a fleshy underground stem modification that has storage parenchyma. Owing to its origin from the stem, a rhizome has nodes and internodes as well as axillary and terminal buds. Adventitious roots develop from underground internodes. The leaves on the rhizome are membranous, dry, and scaly. Rhizomes are brown in color due to a lack of sunlight and often get mistaken for roots such as ginger. When the underground stems are short, swollen, and vertically compressed, but otherwise similar to rhizomes, we call them **corms**. You can see this feature in *Gladiolus*. A corm consists of several internodes and one or two growing apical buds. Adventitious roots develop from a basal node. Superficially, a corm resembles a bulb but remains differentiated from a bulb in having a solid storage tissue inside. **Bulbs** are shortened and compressed underground stems with one or more buds surrounded by numerous fleshy leaf bases that form layers. Fibrous roots develop from the base of the stem, and the apical and axillary buds sprout into aerial stems. Onions and tulips are examples of bulbs.

Tubers are swollen tips of either stem extensions or storage roots. Root tubers are found in sweet potato and cassava. They contain enormous storage tissue inside. On the other hand, stem tubers such as the white potato develop from slender, branched, underground stem extensions called stolons. They have nodes and internodes and may develop adventitious roots at the bottom and new plants from the axillary buds. When the tips of stolons enlarge due to food storage, they form stem tubers. The "eyes" on the white potato are the nodal portions with axillary buds.

Many of the modifications described above provide us with vegetables and staple food items. Some important ones follow.

We know cassava, *Manihot esculenta*, by such other names as **manioc**, **tapioca**, and **yuca**. Cassava plants are tall shrubby perennials with distinct, palmately compound leaves and a cluster of underground storage roots. Cassava is the staple food item for millions of people in the tropical countries, especially in South America and Africa. Starch and milky latex fill the large, massive, brown, thick-skinned storage roots. The latex contains poisonous **cyanogenic glycosides** that we must remove for human consumption. The sweet maniocs contain very few glycosides compared to the bitter varieties. Since it is hard to distinguish the two kinds, people grate and pound both varieties to remove the latex then use the remaining wet mass to make a flatbread. Or, they can dry and toast it to make a flour called **farofa** that keeps fresh for several months. Sizing starch, crisps, fermented beverages, and tapioca are other products from cassava. Tapioca is gluten-free and used in gluten-free baking mixes as a thickening agent. In Africa, the leading producer of cassava, they boil and dry the roots, either grating them later to make a flour or re-boiling them to make a starchy food item. In Ecuador and Colombia people know cassava by the name yuca. In Indonesia, they boil, fry, or bake cassava or make it into a sweet dessert. In India and adjoining countries, besides eating the boiled roots, they produce a granular, white, pearly **sabudana** that they use in place of a similar product called **sago** obtained from the pith of sago palm, *Metroxylon sagu*.

White potato or Irish potato, *Solanum tuberosum*, is a perennial plant from Solanaceae that includes tomatoes and bell peppers. The underground stem tubers rank as the fourth major staple food. Potatoes originated in the Andes Mountains of South America. Archaeological evidence suggests that people ate wild potatoes thirteen thousand years ago in Chile. Native Indians cultivated potatoes extensively all along the west coast of South America and used them as a major food item. The processing methods to preserve the potato starch utilize the weather pattern of the mountains: freezing nights and warm days. To keep the potato starch for long periods of time, they stomp the frozen potatoes to squeeze out the starch. They then allow the cellulose-starch mass to dry. This mix, when dried, yields a mold-tasting powder called **chuno** that people can use for months. A modern freeze-drying process is more rapid and efficient, unlike traditional methods.

Spanish Conquistadors introduced potatoes to Spain in the mid sixteenth century. By the eighteenth century, most parts of Europe accepted the potato as a valuable food item. Ireland was one of the countries where potatoes gained popularity due to suitable soil and climate conditions. It became an important food item in the majority of households by the middle of the 1800s. As the entire country was getting dependent on potatoes, ***Phytophthora infestans***, a fungal pathogen, caused a disease called **potato blight** that intermittently wiped out entire crop fields in a period of five years. This led to the Great Irish famine of 1845–1849 that resulted in the deaths of millions of local people and the emigration of others to foreign countries, mainly to the United States.

Cultivation of potatoes takes place throughout the temperate parts of the world and in elevated areas of the tropics where the temperature is suitable for the crop. There are about five thousand potato varieties around

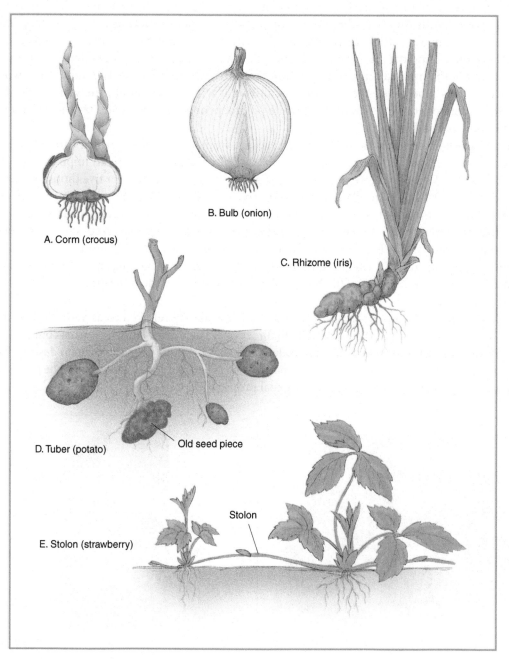

Fig. 3.5 Stem modifications
© Kendall/Hunt Publishing Company

the world, the majority of them in the Andes Mountains. Yukon gold and Peruvian purple are some of the modern cultivars of *Solanum tuberosum*. Around the world, people bake, boil, or fry potatoes with or without their skins and eat them plain or with spices. In Italy, the Italians prepare a pasta called **gnocchi** from potato. Even though the fruits and leaves of potato plants contain poisonous alkaloids, cooked potato is completely safe for human consumption.

Sweet potatoes, **Ipomoea batatas**, are from the morning glory family. The plant is a vine that produces adventitious roots along the trailing stem and large, attractive flowers. Depending on the climate, they can be annual or perennial. The storage roots contain starch, sugar, little protein, vitamin C, and a modest amount of beta-carotene found in the orange-fleshed varieties rather than in the white-fleshed ones. The tubers are often referred to as yams in the United States; true yams are from a monocot plant. Sweet potatoes originated in South America where the population grew them widely as a crop for several thousand years before the European settlers arrived. Several Pacific islands and New Zealand have records of sweet potato cultivation around the same time. Whether the species originated independently in these places or migrated from South America remains a question that intrigues archaeologists. China is the world's leading producer of this tuber, even though people cultivate it widely throughout the world, mainly through cuttings. Sweet potatoes cook very much like white potato; people bake, boil, and fry them. Other products include starch, wine, and alcohol.

Yams, **Dioscorea rotundata** from Dioscoreaceae, are from a vine widely grown throughout the world. Starchy tubers vary in size and superficially resemble a white potato. Yams are an important staple in their land of native origin, West Africa, where the natives have cultivated them for over five thousand years. In New Guinea and adjacent islands yams play an important role in traditional rituals, such as yam cults, where the size of the yams reflects the status of the grower. Growers cultivate yams asexually using small pieces of the sprouting tubers. Yams contain a relatively higher percentage of protein compared to other tubers. Besides fibers, vitamin C, and vitamin B6, they also contain oxalic acid, which accumulates just below the skin area. Peeling and boiling the tubers is essential when cooking these tubers. The presence of sapogenic glycosides in yams has led to the development of birth control pills. You can store the tubers for several months at room temperature.

Taro, **Colocasia esculenta**, is a corm from Araceae. Mistakenly referred to as yam, taro seems to have originated in Southeast Asia and then spread in all directions. The brown-skinned corms contain starch, little protein, and calcium that comes from calcium oxalate crystals. One must cook the corm to destroy the crystals. In Hawaii, the mashed corms make a fermented native dish called **poi**. They can process the starch into flour. One may use the apical buds of the corms for asexual propagation.

Fig. 3.6 Taro
© Lotus Images/Shutterstock.com

Fig. 3.7 Garlic
© Emin Ozkan/Shutterstock.com

TABLE 3.2 Vegetables derived from modified stems and roots

Family	Vegetable Name	Binomial	Modification	Origin
Alliaceae	Onion	*Allium cepa*	Bulb	Asia
Araceae	Taro	*Colocasia esculenta*	Corm	Asia
Convolvulaceae	Sweet potato	*Ipomoea batatas*	Tuber	South America
Dioscoreaceae	Yam	*Dioscorea rotundata*	Tuber	Africa
Euphorbiaceae	Cassava	*Manihot esculenta*	Storage roots	South America
Solanaceae	White potato	*Solanum tuberosum*	Tuber	South America

Onions, **Allium cepa**, and their relatives, garlic, shallots, leeks, and chives, belong to the genus *Allium* and are part of a monocot family Alliaceae. Onions were one of the oldest vegetables used in the world's culinary practices. Onions and garlic are native to Asia, but Egyptians cultivated them by 3200 BCE. Onions and shallots have a single, large bulb, whereas garlic has a cluster of bulbs, called the cloves. The pungent quality of the alliums is due to volatile sulfur compounds that dissolve in water to become sulfuric acid. Onions come in a range of flavors ranging from mild to pungent, and one can add them raw or cooked to other foods. Growers vegetatively propagate them from tiny bulbs called sets. Recipes around the world very much favor garlic as a main ingredient except in the United States where its use is restricted as a flavoring agent to items like garlic bread and garlic butter. Due to its blood thinning properties people have used garlic in medicine since ancient times. It is a rich source of selenium that researchers think has some control on cancer-causing agents and in controlling prostate cancer.

Vegetable Derived from Flower Parts

Artichoke, **Cynara scolymus**, is included in the sunflower family, Asteraceae. It is a perennial plant with a massive inflorescence called the capitulum or the head. People use the immature heads as vegetables. The immature florets and intermingled chaff form the "choke," and the fleshy lower portions of the green, the leathery involucral bracts, and the receptacle form the "heart" of the artichoke. Artichoke in Arabic language means "earth thorn."

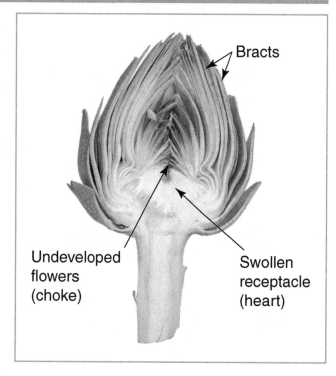

Fig. 3.8 Artichoke
© AN NGUYEN/Shutterstock.com

Review Questions

Give brief answers in the given space.

1. List all cruciferous vegetables discussed in this chapter.

2. What are hypocotyl vegetables? Name any two.

3. Cauliflower is not a real flower. Why?

4. List all vegetables derived from Amaranthaceae.

5. Describe the similarities and differences between cabbage and lettuce.

6. Compare and contrast a rhizome and a corm using one example for each.

7. Discuss the tubers that we use as staple foods.

8. Explain the processing involved in cassava preparation.

9. What are the historical events behind the "Irish famine"?

10. What are true yams? Why are they not suitable uncooked?

Multiple Choice Questions

For each question, choose one best answer.

1. Specialized stems that are underground:

    A. lack nodes, internodes, and axillary buds.
    B. store food material.
    C. are used for vegetative propagation of plants.
    D. all of the above.
    E. only B and C.

2. Which of the following commercially sold plant parts is a hypocotyl vegetable?

    A. Kohlrabi
    B. Ginger
    C. Turnip
    D. All of the above
    E. Both B and C

3. The sweet potato is actually a

    A. rhizome.
    B. tap root.
    C. tuberous root.
    D. stem tuber.
    E. corm.

4. Cyanogenic glycosides occur in the uncooked parts of:

    A. radishes.
    B. yams.
    C. cassava.
    D. taro.
    E. sweet potatoes.

Match the vegetable genera from Column 1 to their family in Column 2. You may use a choice more than once or not at all.

Column 1	Column 2
5. Artichoke	A. Amaranthaceae
	B. Asteraceae
6. Manihot	C. Apiaceae
	D. Euphorbiaceae
7. Lactuca	E. Solanaceae

8. The vegetable part of *Asparagus* is _____.
    A. unexpanded shoots
    B. swollen stem base
    C. leaves
    D. axillary buds
    E. unopened flowers

9. A mold-tasting starch powder derived from white potato is _____.
    A. tapioca
    B. chuno
    C. farofa
    D. all of the above
    E. both A and C

10. Celery and carrots have which of the following in common?
    A. Genus
    B. Family
    C. Species
    D. All of the above
    E. Only A and B

True/false Questions

1. In *Brassica oleracea,* the species name translates to "most common."   **True**   **False**

2. Radish and turnip are included in the same botanical family.   **True**   **False**

3. When bolting occurs, the cabbage becomes useless as a vegetable.   **True**   **False**

4. In the artichoke the vegetable part is the unexpanded stem.   **True**   **False**

5. Farofa derives from sweet potato starch.   **True**   **False**

6. Daikon is also known as Japanese radish.   **True**   **False**

7. Originally carrots were purple in color.   **True**   **False**

8. The edible part of the onion bulb is the fleshy leaf bases.   **True**   **False**

9. Calcium oxalate occurs in the tubers of the white potato.   **True**   **False**

10. Poi is derived from true yams.   **True**   **False**

CHAPTER 4
FRUITS

If a classroom exercise solicited a list of fruits, most students would come up with names such as apple, orange, banana, peach, and grapes rather than beans, okra, corn, and coconut. Yet every single one mentioned above and many more that you would normally call nuts or vegetables are, in fact, fruits. The structure known as **fruit** occurs only in angiosperms. A fruit is technically a mature ovary and its contents; the process of fertilization in a flower initiates the fruit formation.

The ovary becomes the fruit; the ovary wall becomes the fruit wall (also called **pericarp**); and ovules become the seeds. However, there are three possibilities in making the final product that we call a "fruit." A fruit may develop (1) solely from the fertilized ovary; (2) from an unfertilized but mature ovary (**parthenocarpic fruit**); and (3) from the ovary (fertilized or not) *plus* additional parts of the flower, such as the receptacle, petals, and sepals (**accessory fruit**). At maturity, a fruit may have fleshy tissue or it may become dry. When dry, it may split open or not open up by itself. To understand the complexity in the botanical classification of fruits, let us focus on basic details regarding a typical fruit structure. We will use avocado fruit as an example.

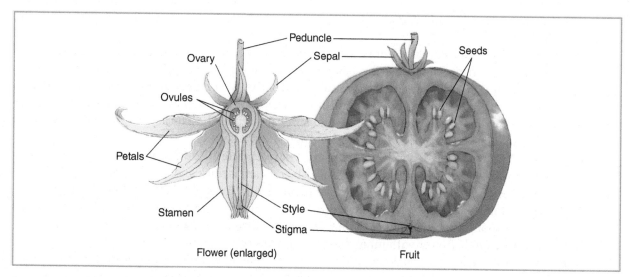

Fig. 4.1 Changes in a fertilized flower
© Kendall/Hunt Publishing Company

51

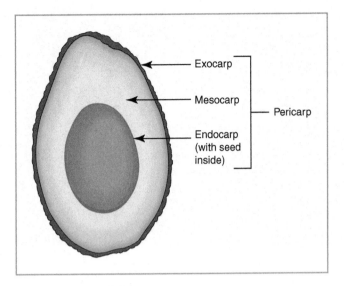

Fig. 4.2 A typical fruit
© Marie Appert/Shutterstock.com

A large, single seed is surrounded by the pericarp that has three layers: an outermost, green layer called the **exocarp** (the skin of the fruit); a yellowish green fleshy layer called the **mesocarp** (the middle layer); and a thin, brown layer just outside of the seed called the **endocarp** (the inner layer). The avocado has a single large seed, but the number of seeds in other fruits may vary. Basically, every single angiosperm fruit ranging from watermelon to chestnut has these three layers even though they may not be readily visible. The variations in texture, color, and thickness of the pericarp result in an assortment of fruits that are distinct for each species. This should explain why you need a knife to cut through a cantaloupe whereas blueberries can go straight into your mouth without all that hassle.

There are two major categories of fruits based on the pericarp.

Fleshy fruits: The entire pericarp or any one or more of the three layers remain fleshy, resulting in a fleshy fruit.

Dry fruits: The entire pericarp is dry at the time of maturity.

The following is a basic classification of fruits under each of the above categories. It lists only the most common fruit types.

Fleshy Fruits

1. **Derived from a single flower with one ovary, resulting in one fruit**
   A. **Berry**—The berry develops from either a superior or an inferior ovary. The exocarp is thin; both the mesocarp and endocarp are fused and fleshy; and seeds are numerous. *Examples:* Tomato, Blueberry.
   B. **Drupe**—The drupe develops from a superior or rarely inferior ovary. Botanists commonly refer to them as "stone fruits," due to having a pit inside. The exocarp is relatively thin, and the mesocarp is either fleshy or fibrous. The endocarp is stony and encloses a single seed. *Examples:* Peach, Mango.
   C. **Pome**—The pome develops from a semi-inferior ovary. The receptacle that surrounds the ovary also becomes part of the fruit, making it an accessory fruit. The exocarp is thin; the mesocarp is fleshy; and the endocarp is thin. *Examples:* Apple, Pear.
   D. **Pepo**—A pepo develops from an inferior ovary. The receptacle that surrounds the ovary also becomes part of the fruit, making it an accessory fruit. The exocarp is thick and forms a hard rind; the mesocarp and endocarp are fleshy, and the seeds are numerous. *Examples:* Watermelon, Pumpkin.
   E. **Hesperidium**—The hesperidium develops from a superior ovary. The exocarp is leathery, the mesocarp spongy, and the endocarp thin, with juice-filled hairs. *Examples:* Orange, Grapefruit.
2. **Derived from a single flower with many separate ovaries resulting in one fruit**
   Aggregate fruit—This is a conglomeration of several individual units made of berries, drupes, or achenes that make up one single fruit. The ovary position is not critical, even though most known fruits are from superior ovaries. *Examples:* Strawberry, Blackberry.

TABLE 4.1 Fleshy fruits

Fruit Derived From	Fruit Type	Most Common Ovary Position	Example	Special Note
Single flower, single ovary	Berry	Superior or Inferior	Guava, Kiwi, Cranberry	Fleshy; some dry out eventually
	Drupe	Superior or inferior	Cherry, Olive, Coconut	Mesocarp fleshy to fibrous; some dry out and resemble a nut
	Pome	Semi-inferior	Quince, Pear	Accessory tissue
	Pepo	Inferior	All types of squashes, melons	Accessory tissue; thick exocarp
	Hesperidium	Superior	All citrus fruits	Leathery exocarp
Single flower; many ovaries	Aggregate fruit	Superior on a common receptacle	Raspberry, Custard apple	Some dry fruits like Magnolia are aggregate
Many flowers of an inflorescence	Multiple fruits	Superior or inferior	Mulberry, Noni	Accessory tissue; various shapes based on inflorescence type

3. **Derived from multiple flowers of an inflorescence resulting in one fruit**
   Multiple fruit—This is a single fruit but formed as a result of the union of the mature ovaries, the surrounding flower parts, the receptacle, and the central axis of the inflorescence. These are invariably accessory fruits. The ovary position is not critical, and we find individual flowers with either a superior or an inferior ovary in some known fruits. *Examples:* Pineapple, Figs.

Dry Fruits

There are two kinds of dry fruits. Irrespective of the development from inferior or superior ovaries, botanists classify these into the following:

1. **Dry, dehiscent fruits: The pericarp will split open to release the seeds for dispersal.**
   A. **Legume**—At maturity, the fruits split open in two sutures, dorsal and ventral. *Example:* Beans.
   B. **Follicle**—At maturity, the fruits split along one suture. *Example:* Milkweed.
   C. **Capsule**—At maturity, the fruits split open in more than two sutures. This is the most abundant dry, dehiscent fruit. The capsules split open in a variety of ways. *Examples:* Okra, Poppy.
   D. **Silique**—At maturity, it splits open in two sutures, except that the seeds attach in a partition called a replum. *Example:* Mustard.
2. **Dry, indehiscent fruits: The pericarp will not split open but moves along with the seeds during dispersal.**
   A. **Achene**—A single-seeded fruit with a dry, loose pericarp. *Examples:* Sunflower, buttercup.
   B. **Caryopsis or Grain**—A single-seeded fruit with the pericarp firmly fused with the seed coat. *Examples:* Rice, Wheat.
   C. **Nut**—A single-seeded fruit with the entire pericarp forming a hard, stony, loose layer surrounding the seed. *Examples:* Chestnut, Acorn.

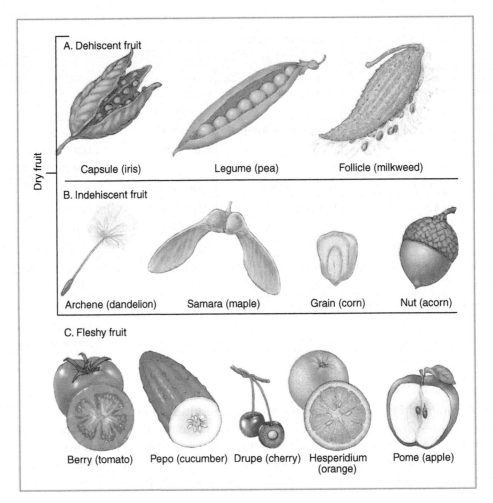

Fig. 4.3 Major fruit types
© Kendall/Hunt Publishing Company

D. **Samara**—A modified achene, with the pericarp developing into a wing-like structure that aids in dispersal. *Examples:* Elm, Maple.
E. **Schizocarp**—Composed of two units, each with its own single seed called a mericarp, which separate easily from each other in the middle. *Examples:* Cumin, Dill.

Some Exceptions

Most fruits that we will learn about in this chapter safely fall into one of the above categories. But there are exceptions to this. We may not readily categorize some fruits due to their complex nature. For example, the banana is a pseudo-berry; the avocado is a drupe without a pit, or a single-seeded berry; and pomegranates and passion fruits are exceptional berries. Litchi is a drupe, even though what appears to be the mesocarp is a fleshy seed coat, and the papaya is a pepo derived from a superior ovary. Rose hips are accessory fruits that are either dry or partially fleshy. We will address other notable exceptions under each individual fruit. As one may already realize, there are hundreds of edible fruits around the world. It is impossible to discuss every one of them due to limitations in this text. The sample of fruits covered here concern those fruits that commonly show up in fruit markets around most parts of the world.

Some angiosperm families, more than others, provide us with a number of edible fruits. Under fruit type we will focus on these families first, in no particular order, followed by less represented families.

Fig. 4.4 Aggregate and multiple fruits

Soursop: © COLOA Studio/Shutterstock.com; Rose hips: © pikselstock/Shutterstock.com; Fig: © azure1/Shutterstock.com; Mulberry: © Madlen/Shutterstock.com; Strawberry: © Tim UR/Shutterstock.com; Blackberry: © Dionisvera/Shutterstock.com; Pineapple: © Viktar Malyshchyts/Shutterstock.com

TABLE 4.2 Dry fruits

Dehiscent/ Indehiscent	Fruit Type	Nature of Pericarp	Example
Dehiscent	Legume	Opens in two sutures	All kinds of beans and peas
	Follicle	Opens in one suture	Magnolia, Milkweed
	Capsule	Opens variously	Vanilla, Okra
	Silique	Opens in two sutures; replum present	Most members of Mustard family
Indehiscent	Achene	Thin and loose around a single seed	Most members of Amaranth family
	Caryposis	Fused with seed coat	All grains
	Nut	Thick and loose around a single seed	Hazelnut
	Samara	Develops into winged structure around the seed	Most members of Elm and Maple family
	Schizocarp	Composed of two units; each split into single seeded units with its portion of pericarp	Most members of Carrot family

Rose Family Fruits

We also refer to the rose family as **Rosaceae**. The family gives us roses, cherry blossoms, and a number of fruits that we cannot bypass. Botanists traditionally subdivided this large family with its wide variety of fruits into four smaller groups, with the fruit type as a main criterion.

> **Subfamily Maloideae:** includes pome fruits like apple and pear
>
> **Subfamily Amygdaloideae:** includes stone fruits like peaches and cherries
>
> **Subfamily Rosoideae:** includes aggregate fruits like strawberry and rose hips
>
> **Subfamily Spiraeoideae:** includes fruits that are dry and inedible like bridal wreath

Pome Fruits—Subfamily Maloideae

Apples, *Malus pumila,* are one of the oldest fruits associated with human history, and people have cultivated them for a long time. The trees are medium-sized and perennial and can bear fruit for up to a century. Apple blossoms are fragrant, pollinated by bees. To ensure uniformity in taste and to promote ease of harvest, farmers use grafting (a form of asexual reproduction) to obtain quicker and more desirable results. There are thousands of varieties of apples but only a few like Red Delicious, Golden Delicious, Granny Smith, Jonathan, McIntosh, and Rome. These are the varieties most commonly sold in supermarkets.

In Greek mythology, apples were the symbol of immortality, knowledge, and temptation. *Malus* in Latin means "bad," referring to Adam and Eve's fateful consequences after eating the fruit. Controversies exist about the identity of the fruit mentioned in the Bible due to the fact that apples did not grow in the Mediterranean region where the story supposedly took place.

Wild apples probably were native to Asia where people ate them as fruit since prehistoric times. The cultivated ones are all in a group with one name, *Malus domesticus*. In the United States, the colonists first planted them in Bunker Hill in the 1620s. European colonists brought the grafting techniques to North America and established grafted trees in the apple orchards. John Chapman of Massachusetts, popularly known as "Johnny Appleseed" (1774–1842), helped spread the trees through seedlings raised from the seeds. He raised apple trees in nurseries that he owned in Ohio, Pennsylvania, Kentucky, Illinois, and Indiana. Johnny also supplied

TABLE 4.3 Subfamilies of rosaceae

Sub Family	Fruit Type	Example	Binomial	Origin
Maloideae	Pome	Apple	*Malus pumila*	Central Asia
		Pear	*Pyrus communis*	Asia
		Quince	*Cydonia oblonga*	Eurasia
Amygdaloideae	Drupe	Peach	*Prunus persica*	Western China
		Plum	*Prunus domestica*	Asia
		Apricot	*Prunus armeniaca*	Asia
		Cherry—sweet	*Prunus avium*	Near East
		Cherry—sour	*Prunus cerasus*	Near East
Rosoideae	Aggregate fruits	Virginia Strawberry	*Fragoria virginiana*	North America
		Blackberry	*Rubus fruticosus*	North temperate regions
		Raspberry-black	*Rubus occidentalis*	North temperate regions
		Raspberry-red	*Rubus idaeus*	North temperate regions
Spiraeoideae	Capsules; Follicles	Meadowsweet	*Spiraea ulmaria*	North temperate regions

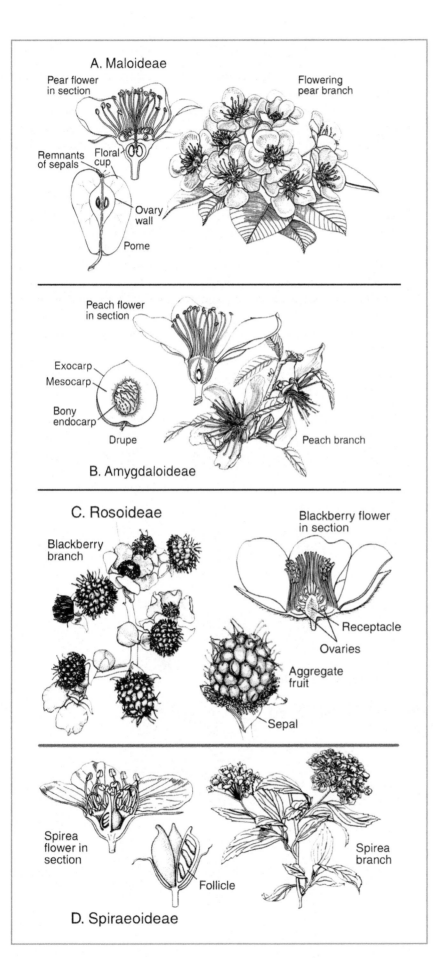

A. Maloideae

Pear flower in section

Flowering pear branch

Remnants of sepals

Floral cup

Ovary wall

Pome

B. Amygdaloideae

Peach flower in section

Exocarp

Mesocarp

Bony endocarp

Drupe

Peach branch

C. Rosoideae

Blackberry branch

Blackberry flower in section

Receptacle

Ovaries

Aggregate fruit

Sepal

D. Spiraeoideae

Spirea flower in section

Follicle

Spirea branch

Fig. 4.5 Four subfamilies of Rosaceae are illustrated to show the floral and fruit characteristics

From *Economic Botany: Plants in Our World* 3E by Beryl Simpson & Molly Conner Ogorzaly, © 2001. Reproduced by permission of The McGraw-Hill Companies.

apple seeds to the pioneers in the mid-western United States. Apples raised from seeds eventually resulted in a variation in American apples from the grafted European ones.

The accessory tissues in apples, pears, and quinces derive from the receptacle surrounding the semi-inferior ovary. Pears are the second largest temperate fruit crop in the world. The gritty texture of pear fruit is due in part to **stone cells** or sclereids found among soft parenchyma cells. **Quinces**, *Cydonia oblonga,* often mistaken for pears except for their sweeter, aromatic flesh, are popular in Europe.

Apple Facts

- The Geneva Respository in New York maintains about 2,500 apple varieties from all over the world.
- Despite the close association of apples with American life, apples originated in the Old World.
- American Red and Yellow Delicious varieties are the most popular in the commercial world.
- Washington apples remain number one among the American public.
- On the average, a consumer in the United States eats about nineteen pounds of fresh apples a year, i.e., one apple per week.
- With the health benefits provided by quercetin, pectin, and tannin, an apple a day may keep the doctor away.

Stone Fruits—Subfamily Amygdaloideae

Fruits with a hard endocarp (stone) enclosing a single seed called a "pit" exist in all members of this subfamily. Peaches, nectarines, apricots, plums, and cherries are some stone fruits. Growers graft most of these like apple trees. For binomials and their origin, please refer to Table 4.3. Almonds are in the same group, which we discuss later in the chapter.

Peaches are the third most important fruit crop in the United States. For several thousand years people in China have eaten them as fruit. Spaniards brought this fruit to Mexico in the 1500s. The United States is the

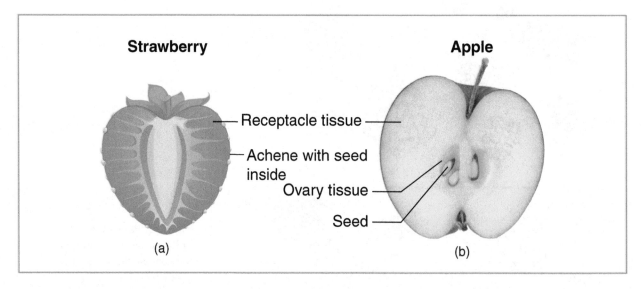

Fig. 4.6 Accessory tissues compared

(a) © Kozhadub Sergei/Shutterstock.com
(b) © photomaster.com

leading producer with orchards found in California, Texas, Arkansas, New Jersey, and Georgia, the peach state. People prefer freestone peaches for canning rather than the cling type. Nectarines are a recent cultivar group of peach that started as a mutant from peaches. Nectarines lack the furry skin of peaches.

Apricots have a distinctively bitter flavor compared to other stone fruits in this group. In ancient China, people associated eating apricots with prophetic powers. In Persia, dried apricots were an important item of commodity for trade. Persians called them "seeds of the sun." In Hunza, a small kingdom in the Himalayas, the local people attribute their long healthy lives to including apricots in their diet. Most American apricots came from those established by Spaniards from Greece where they arrived from Persia through Alexander the Great. Turkey is the world's leading producer of apricots, and within the United States, California leads in apricot cultivation. Pluots are the modern crosses between the apricot and the plum.

Plums come from a variety of species widely cultivated around the world's temperate regions, ranging from Europe, Western and Central Asia, and North America. The prune is the dried plum that consumers commonly use as a natural laxative. You can buy pickled plums in Asian specialty stores.

Cherries are of two main types: the sweet cherry and the sour cherry. Each has several cultivars. One cannot readily distinguish these two types though except for their taste and slight difference in size. Growers sell sweet cherries for consumption. They make maraschino cherries from the Royal Anne variety, a sweet type that they have bleached, deseeded, and soaked in sugar solution with added dye and flavoring. California and Washington grow most of the sweet cherries within the United States. Sour cherries are useful in pies, salads, jams, and jellies. The leading states in the production of sour cherries are Michigan, New York, and Utah. Cherry trees became established in the United States by the year 1847, and since then people have preferred them for their flowers rather than for their fruits. The presence of anthocyanins in cherries has created an interest in medical research about their potential benefits for reducing pain and inflammation.

Aggregate Fruits—Subfamily Rosoideae

Fruits included in this subfamily are aggregate fruits that include strawberry, blackberry, and raspberry. These fruits derive from a single flower with several separate pistils seated on a common receptacle, each with a superior ovary. The shape of the receptacle may vary from a flat disc to a conical structure. The fleshy portion of the fruit may derive from different parts of the flower. In blackberry, raspberry, and similar plants commonly referred to as **brambles**, each fertilized ovary ripens into a drupelet and adheres with the adjacent drupelet and so on to form a conglomeration of individual fleshy drupelets. What we refer to as a single fruit is an aggregate fruit with each tiny sphere on it presenting an individual drupe with its own pit (hard endocarp) inside. The receptacle in these fruits does not become fleshy and either remains free or adheres to the rest of the fruit.

A number of species of a single genus **Rubus** give us blackberries, raspberries, and several hybrids, including loganberries. An edible purple dye comes from the fruits of black raspberries that USDA rating stamps use on meats.

Strawberries are aggregate fruits with individual ovaries that ripen into single-seeded achenes. These achenes appear as brown speckles scattered on the surface of the fruit. The fleshy portion of the fruit is the accessory tissue of the receptacle. The genus name **Fragaria** derive from the Latin word *frago* (= fragrant), referring to the sweet, delightful scent of the fruit. People have eaten strawberries since ancient times throughout Europe and the Americas. There are several species, including the natural hybrid that originated from Chilean and Virginia species in the 1750s. Modern strawberries are cultivars developed for more fruit yield in contrast to wild species. The United States leads as the single most country in the cultivation of strawberries.

Dry Fruits—Subfamily Spiraeoideae

Fruits of this subfamily are dry and inedible to humans. Consumers use most members as ornamentals.

Cucurbit Family Fruits

We know this family as **Cucurbitaceae** and its fruits as squashes and melons. Ancient human history has known them around the world. Squashes belong to the genus *Cucurbita* that includes all of the winter and summer squashes, including zucchini and pumpkin. For the native people of Mexico, South America, and North America, squashes were one of the main items of food. They called them one of the "three sisters" that included corn and beans. *Cucurbita pepo* is the most widely known and cultivated species in all tropical regions of the world. Archaeological evidences from Mexico suggest the use of cucurbits since 6000 BCE.

The plants of this family are annuals, raised from seeds, and they grow into a creeper or climber with unisexual flowers. Most cultivated species are monoecious even though dioecious plants are common in the wild. Consumers use the leaves and blossoms in specialty soups and sauces. The most useful part of the plants is the fruit called **pepo**, which derives from the inferior ovary. The exocarp is a hard rind derived from the ovary wall, and the accessory tissue derives from the receptacle. The mesocarp and endocarp are fleshy and indistinguishable and enclose numerous, flattened seeds. Both the flesh and the seeds of many squashes are rich in sulfur-containing amino acids. Unripe fruits are useful as vegetables (please see fruit or vegetable); we describe the ones eaten as fruits below.

Cantaloupe, *Cucumis melo,* is an edible fruit from Cucurbitaceae. Native to India and Africa, people cultivate it widely elsewhere in the tropics. Fruits are round to oblong, about ten inches long, with a thick, greenish-brown, rough rind. The flesh is yellow-orange, sweet and juicy. It contains polyphenol antioxidants, known for promoting a healthy cardiovascular and immune system.

Honeydew is a cultivar of *Cucumis melo.* The fruits are round to oval shape, about the size of cantaloupe. The rind is yellow green and smooth. The flesh is green, sweet and juicy, and a rich source of vitamin C and sugars.

Watermelon is *Citrullus lanatus*, a native of Africa cultivated in the adjacent tropics since ancient times. China is the leading producer of watermelons where they have existed since the tenth century CE. Native Americans cultivated New World watermelons even before the French settlers arrived. European colonists and African slaves might have played a key role in spreading watermelons throughout the world. Within the United States, Texas and California cultivate the most watermelons. It became the state vegetable of Oklahoma in 2007; however there is controversy about it being a fruit. In China, the Chinese offer watermelon seeds dyed red to friends and family during New Year celebrations as a symbol of truth, joy, happiness, and sincerity.

The fruit is huge, heavy, and round or oblong with a weight ranging from about eight to twenty-five pounds. The thick rind is smooth and green and usually mottled in dark green or yellow. The flesh is red, juicy, and sweet with numerous black seeds intermingled with placental tissue. Newer varieties include the

TABLE 4.4 Citrus fruits

Common Name	Binomial	Main Use	Notes
Bitter orange	*Citrus aurantium*	In marmalade, liqueurs; perfumes	Hybrid origin
Citron	*Citrus medica*	Fruit peel in confections	Used in Persia since 4000 BCE.
Grapefruit	*Citrus paradisi*	Fruit	Hybrid origin; drug interaction
Lemon	*Citrus limon*	Perfumes; deodorants; in cleaning products	Hybrid origin
Mandarin orange/ Tangerine	*Citrus reticulata*	Fruits with loose skin; eaten fresh or for juice	Natural distinct species
Mexican lime/ West Indian lime	*Citrus aurantifolia*	In cosmetic products	Hybrid origin; high vitamin C content
Pummelo	*Citrus maxima*	Large fruit, thick rind; eaten fresh or used in preserves	Popular in Thailand and most of Asia
Sweet orange	*Citrus sinensis*	Fruit	Most widely grown

seedless kinds, the yellow flesh types, and smaller fruits about the size of honeydew. Watermelons are high in acidity and water content as well as in vitamins A and C.

Citrus Family Fruits

Citrus fruits are hesperidia, and all of them belong to one single family, **Rutaceae**. Oranges, lemons, limes, tangerines, citrons, and grapefruits belong to the genus *Citrus*; kumquats are in a related genus *Fortunella*. The fruit has a leathery, thick exocarp with numerous glandular pockets filled with aromatic oil that gives off the flavor characteristic of the citrus fruits. The mesocarp is the spongy white layer that adheres to the exocarp. The endocarp has a modification of fleshy hairs filled with citrus juice. Citrus plants originated in tropical southeast Asia. Oranges and tangerines play an important role in Chinese holiday traditions and represent abundant happiness. Many cultivated species

© D7INAMI7S/Shutterstock.com

cannot tolerate a severe freeze but thrive well under sunlight and abundant rainfall. Growers graft most trees, although they can raise a few from the seeds. Table 4.4 lists some of the well-known citrus fruits around the world.

Heath Family Fruits

Two native fruits of North America, blueberries and cranberries, are from **Ericaceae**, a family that includes azaleas used in subtropical landscaping. Blueberry and cranberry plants thrive well in acidic soil. A circular scar or a crown of calyx that remains on top of the berries indicates the fruits' origins from an inferior ovary.

Fig. 4.7 Blueberries
© Maria Dryfhout/Shutterstock.com

Blueberries derive from any one of three species of *Vaccinium*, as Table 4.5 shows. There are numerous cultivars of high-bush blueberry now grown in Europe, South Africa, South America, and Australia. Within the United States, Maine and New Jersey are the leading producers of blueberries. The plant is a shrub varying in size from a low bush to a high bush of about four meters tall. The fruits are small, about one inch in diameter or smaller, and they are deep purple in color with a dull, sweet taste and variable acidity. Consumers eat blueberries as fresh fruits or use them in jams, beverages, pancakes, muffins, and pies. Scientists have discovered their value as "antioxidant powerhouses" recently, and since then have promoted them as one of the **superfruits**. Fruits that belong to this new nutrient category are loaded with favorable phytonutrients and antioxidants.

The genus *Vaccinium* also includes cranberry plants. These are creeping shrubs grown in frequently flooded moist, acid peat bogs. The edible berries are about the size of blueberries, red and tart. Cranberry picking involves simply shaking the branches to loosen the fruits that fall and float on the flooded water. Large machinery then scoops them out. Another method of harvesting is dry harvesting where a mechanical picker with a large comb-like attachment separates fruits placed in a conveyor that eventually collects them in a sack. Cranberry sauce is an important menu item in North American and Canadian Thanksgiving tradition. People serve it as a relish along with baked turkey. Native Americans used the fruits fresh or mashed with cornmeal and also obtained medicinal teas and dyes from the berries. Consumers commonly eat berries fresh or in sauces, relishes, jellies, and beverages. Cranberry juice is a popular cure for urinary tract infections. The fruits are also a rich source of vitamin C, copper, iron, and dietary fiber as well as antioxidants, which give cranberry its **superfruit** status.

Fig Family Fruits

The family **Moraceae** includes figs, jackfruit, breadfruit and mulberry. The trees are medium-sized with sticky latex that is abundant in their tender parts, leaves, and ripe fruits. All of the listed fruits are multiple fruits derived from many flowers of an inflorescence.

TABLE 4.5 Edible fruits of Ericaceae

Common Name	Binomial	Origin
Cranberry	*Vaccinium macrocarpon*	North America
Cranberry	*Vaccinium oxycoccus*	North temperate regions
High-bush blueberry	*Vaccinium corymbosum*	Northern United States
Low-bush blueberry	*Vaccinium angustifolium*	Northern United States
Rabbiteye-blueberry	*Vaccinium ashei*	Southeastern United States

The commercial **fig** is from *Ficus carica,* one of about eight hundred species of a single genus *Ficus* that includes the rubber tree, the banyan tree, and so forth. Figs are native to the Mediterranean region and also grow widely in warmer parts of Africa, Australia, South America, and in the United States. Figs are one of the ancient trees domesticated by humans mentioned in Egyptian documents dating from 2750 BCE. The Bible tells how Adam and Eve used the leaves of figs to cover themselves. Figs are multiple fruits that develop from a syconium, a special inflorescence that has a flask-shaped hollow receptacle with a small opening at the top. Inside of the receptacle there are hundreds of tiny, incomplete, unisexual flowers facing each other. Figs can become either monoecious or dioecious. Wild figs in their native areas become pollinated by tiny gall wasps that enter into the **syconium** through the opening to lay their eggs inside. As they fly in and out, they brush against male flowers and inadvertently dust the pollen on female flowers. In wild figs, wasp eggs and fig seeds intermingle and are hard to distinguish with the naked eye. The recent cultivars grown in the United States do not go through this pollination mechanism because these wasps do not exist here. Such figs do not produce viable seeds even though they ripen into edible parthenocarpic fruits. Growers raise modern figs by cuttings. One can cook figs as a vegetable, eat them fresh or dried, and use them in jam making. Figs are an excellent source of calcium and dietary fiber. We also know them for their laxative effect and high antioxidant levels.

Jackfruit, *Artocarpus heterophyllus,* is an evergreen tree with large leaves. Mature leaves are entire, young leaves are somewhat lobed. The trees are native to the indigenous rain forests of India, and are also widely grown in the tropics of Asia, Africa, and America. The flowers are unisexual and borne on the same tree in separate inflorescences called flower heads. The female inflorescence hangs on thick stalks from the main branch or from the main trunk. Once they ripen, the fruits are huge, heavy, and roughly oblong, measuring about one to three feet long by one to one and a half feet wide. The fruit rind is thick and normally green to pale yellow, with numerous soft, short, pointed outgrowths, each with a dark tip. A musty, sweet, distinct aroma reminiscent of pineapple and banana indicates the fruit's ripeness. When one cuts the fruit open, a central massive core formed from the inflorescence axis or peduncle is visible. Twenty to thirty large brown starchy seeds surround the pithy core, each about one and a half inches long and one inch wide. Enclosing each seed are firm, golden yellow edible flesh and cream-yellow soft fibers (called rags) derived from the perianth and other unfertilized flowers. Consumers eat the flesh surrounding the seeds raw or, as a modern twist, as an accompaniment to ice cream. After removing the brown, brittle skin, they eat the seeds roasted or boiled, or they cook them as a vegetable.

Breadfruit trees, *Artocarpus altilis,* produce similar but smaller fruits than jackfruits. The trees are deciduous in their native areas of the Pacific islands. They have predominantly lobed leaves that are distinct and larger than jackfruit leaves. The fruits are round to oblong, from nine inches to one foot long. The rind is mostly green, fairly smooth, and rough; occasionally it is bluntly prickly. The core and flesh are white to pale yellow, with numerous seeds. Ripe fruits lack the sweetness and aroma of the Jackfruit, and Polynesians use them mostly as a staple. They remove the rind and slice and cook the firm pulp into a bland, starchy dish that resembles rice pudding. They also consume roasted or boiled seeds locally.

Pawpaw Family Fruits

Sops in general refer to several species of **Annona**, **Annonaceae**. Custard apple, soursop, sweetsop, and cherimoya are popular fruits in the tropics of the Americas where they are native. There is a considerable amount of confusion and overlapping in common names. All of these fruits derive from small to medium-sized trees. Fruits are more or less round to ovoid, about the size of a medium apple, and larger in soursops. Several polygonal segments or soft prickly markings on the outside make them readily distinguishable, with each area representing one of several pistils of a single flower. Sops are aggregate fruits with soft, easily separable rinds. Inside there are about twenty to thirty black shiny seeds embedded in soft, white, sweet flesh. Of all known sops, consumers consider cherimoya the most delicious. Different varieties exist with various shapes, textures, flavors, and tastes of the fruits.

(a) Custard apple and Cherimoya—entire and cut view

(b) Jackfruit

(c) Breadfruit

Fig. 4.8 Custard apple, Breadfruit and Jackfruit

(a) © Guan jiangchi/Shutterstock.com; © LianeM/Shutterstock.com
(b) © Le Do/Shutterstock.com; © Ca Dao/Shutterstock.com

TABLE 4.6 Commonly known Sops

Common Name	Binomial	Fruit Description
Sugar apple/ Sweetsop	*Annona squamosa*	More or less round, skin pale grayish green. Polygonal markings are distinct, easily separable. Flesh is sweet and granular.
Soursop	*Annona muricata*	Longer than wide; irregularly shaped; skin dark green covered with soft, thin, prickly outgrowths. Flesh pulpy and softly fibrous
Cherimoya	*Annona cherimoya*	Similar to sugar apple, except for almost smooth green tender skin with faint markings. Flesh is rich and creamy with sweet fruity flavor.
Custard apple	*Annona reticulata*	Roundish, skin dull green to brown. Markings are faint, slightly raised. Flesh is thick, custard-like, and granular.

Fruits from Other Angiosperm Families

Fruits that are Berries

Grapes, *Vitis vinifera,* are in Vitaceae, a climbing vine native to middle Asia. One of the oldest cultivated fruits, known since 4000 BCE. Modern seedless varieties such as "Thompson Seedless" are the cultivars derived from this species. *Vitis labrusca* is the North American native from which growers selected the Concord and Catawba varieties. Commercially grown grapevines are vegetatively propagated by cuttings.

The fruits are rich in simple sugars, and consumers eat them fresh or use them in making jelly, juice, vinegar, wine, grape seed extracts, and grape seed oil. Grape juice is pasteurized to prevent fermentation. This process also alters the delicate taste of most common grape juices. Concord grapes seem to retain the flavor in spite of pasteurization. As a result, Concord grapes have become known as traditional juice grape. Raisins are grapes carefully dried with sulfur dioxide added as a preservative. Muscat of Alexandria, Sultana, and Black Corinth are favored for their soft texture and flavor in raisin making. Grapes are rich in antioxidants, potassium, and vitamin C with moderate amounts of folic acid, calcium, and dietary fiber.

Kiwi, *Actinidia deliciosa,* Actinidiaceae, is native to China. It was later introduced to New Zealand, Italy, South Africa, Chile, and California in the United States where they are still in cultivation. Kiwi plants are vigorous, climbing woody shrubs. The berries are oval, about three inches long, with a dull brown thin skin covered with short, fuzzy hairs. The flesh inside is green, soft, tart, and translucent with a mild, unique flavor. A ring of tiny, black seeds surrounds a central core. Kiwi fruits were originally known by the name "Chinese Gooseberry," which people in New Zealand renamed "kiwi" based on an indigenous flightless bird. Kiwi fruits started to become popular within the United States in the 1980s. Because of this, people refer to them as "the fruits of the eighties." Kiwis are a rich source of fiber, vitamins A and C, and potassium, along with moderate amounts of protein, folic acid, and phosphorous. A protein degrading enzyme called **actinidin** is useful as a meat tenderizer. Fruits are normally eaten fresh or used as garnishes in fruit salads due to their attractive color. Golden kiwifruit is a new cultivar with a sweeter, less acidic yellow flesh that is becoming increasingly popular.

Star fruit, *Averrhoa carambola,* Oxalidaceae, is a greenish-yellow, angular berry that creates a star-like outline when sliced. The trees are medium sized, native to Indonesia, and widely grown in the tropics. The fruits have a sour taste due to a high concentration of calcium oxalate crystals that dissolve in saliva to form oxalic acid. They are a good source of dietary fiber and vitamin C.

Guava, *Psidium guajava,* Myrtaceae, is native to Central America and grows in most tropical areas, including cooler climates. The tree is medium-sized, and the fruits are berries with a peculiar musky, pungent smell. The exocarp may be green or yellow with a light yellow or red pulp inside. The entire fruit is eaten fresh or frequently used in jams, jellies, and tropical punches. It is an excellent source of vitamin C (even more so than citrus fruits), dietary fiber, vitamin A, omega-3 and -6 fatty acids, potassium, magnesium, and antioxidants. Even though it has not yet achieved much popularity, guava is one of the **superfruits**.

Bananas are the most popular tropical fruit, which belong to a monocot family, Musaceae. Bananas originated in eastern Asia and Australia where they were one of the first agricultural crops. They were known in India by 600 BCE. and in the Mediterranean region by 300 BCE. Travelers probably introduced them independently to other warmer parts of the world, including Africa. The Portuguese introduced bananas to South America in the sixteenth century, where they spread to the Central American countries. Since then the banana has remained one of the most important cash crops in Latin American countries, which has led to their being referred to as **banana republics**. All kinds of bananas that we see in fruit markets are from a single genus *Musa*. Hybridization between two diploid species, *Musa acuminata* and *M. balbisiana*, resulted in various degrees of polyploidy, and further domestication yielded numerous kinds of modern bananas grouped under the cultivar name *Musa X paradisiaca*. Brazil is the leading grower of bananas in South America. Costa Rica and Ecuador are the largest exporters of bananas.

Banana trees are actually large herbs without any woody tissue in the stem. Tightly packed petiole sheaths in a cylinder made of overlapped layers form the stem. An average tree can reach twenty feet tall. The glossy

Fig. 4.9 Sample of fruits

Kiwi: © MRS. Siwaporm/Shutterstock.com; Pomegranate: © Valentyn Volkov/Shutterstock.com; Passion fruit: © matin/Shutterstock.com; Banana: © Iurii Kachkovskyi/Shutterstock.com; Litchi: © Anna Kucherova/Shutterstock.com; Rambuttan: © SOMMAI/Shutterstock.com; Starfruit: © Mr. Nakorn/Shutterstock.com; Guava: © Mau Horng/Shutterstock.com; Mango: © Maks Narodenko/Shutterstock.com; Mango core: © aabeele/Shutterstock.com; Durian: © ilovezion/Shutterstock.com; Durian C.S.: © Patchy.nang/Shutterstock.com; Cherry: © Dionisvera/Shutterstock.com; Pummelo: © Cloud7Days/Shutterstock.com

smooth huge leaves are oblong, up to nine feet long by two feet wide. Banana plants propagate asexually through underground rhizomes. The rhizomes sprout into new plants that surround and replace the older trees. The average life span of a banana tree is about two years; they reproduce only once in their life cycle. The inflorescence is a terminal spike with a massive central peduncle that holds several rows of purple colored large bracts in tight spirals. Each bract encloses a bunch of unisexual flowers. The basal eight to fifteen rows are female; the upper rows have male flowers; and there may be some bisexual or neuter flowers in middle rows. The inflorescence bends downward in most plants, opening one or two bracts at a time, starting from the basal most bracts. In a day or two, the open bracts fall off, leaving the female flowers. The entire process takes place anywhere from three weeks to two months until all the bracts fall off except those at the very tip. This forms a conical structure with a combination of the unopened bracts and the last-formed male flowers. These are called "banana flowers" that are removed for use as a vegetable. The banana fruits start to ripen, turning from green to yellow while on trees. Each bunch can have anywhere from ten to twenty-four fruits. The fruit is technically a **pseudo-berry** that lacks viable seeds in cultivars due to parthenocarpy. The dark scar at one end of the banana is the calyx region that indicates the fruit's development from an inferior ovary.

In Asia, people frequently use unripe bananas as vegetables, either frying, boiling, steaming, or roasting them. They also use them in making beer. Green bananas are the basic food for the people of the South Pacific and Caribbean islands. The fruits typically have three angles, and the ripe flesh is creamy white, firm, and slippery with starch and sugar. Bananas are low fat fruits rich in potassium and vitamins C and B6, with a moderate amount of fiber and a small amount of protein. Bananas are well known by those who practice folk medicine. They use all parts in treating bronchitis, dysentery, ulcers, diabetes and digestive disorders. Central Americans use the sap of red bananas as aphrodisiac.

In India, the natives use the banana as a symbol of fertility and prosperity. Every religious and auspicious celebration starts with live banana trees placed at the entryways. They chop banana flowers and use them in cooking. Since the bracts and pistils are indigestible, the cooks remove them from each flower before cooking. They believe the central core of the stem is a cure for kidney stones and take freshly squeezed sap from the stem internally as a home remedy. There are anecdotal evidences for this treatment. Banana leaves are useful for serving and wrapping food, and people use the fiber obtained from the dried petiole sheath to tie flowers in a garland. Chapter 11 discusses the commercial uses of banana fibers.

Passionfruit, *Passiflora edulis,* Passifloraceae, which is native to Brazil, gets cultivated for its fruits and beautiful flowers. People widely cultivate many species in the tropics that yield edible fruits. The fruit is a specialized berry with a tough exocarp that encloses numerous juicy seeds. The juicy flesh is the **aril** (the fleshy seed coat), the edible part of the fruit. The plants are climbing vines with bright, attractive flowers. The flower structure with its fringed corona and gynophore is of interest to botanists. In the fifteenth and sixteenth centuries, Spanish Christian missionaries adopted the passionflower as a symbol of the Crucifixion. The intricate flower parts correspond with the Crucifixion's details: The ten perianth segments represent the ten faithful apostles; the corona represents the Crown of Thorns; the three stigmas represent the three nails; the five anthers represent five wounds. The tendrils on the plant represent the whips. The purple color in the flower represents the color of the robe used to cover Jesus after crowning him. Spanish names such as "Christ's thorn" relate to this symbolism. Fruits are eaten fresh, or used in tropical fruit punch, mousse, ice cream and yogurt. It is a low fat fruit with a good amount of dietary fiber and vitamins A and C.

Fig. 4.10 Passionfruit
© LittleStocker/Shutterstock.com

Pomegranates, *Punica granatum,* Punicaceae, are native to Iran, Afghanistan, Pakistan, and the Himalayan region in northern India. The pomegranate plant ranges from a shrub to a small tree, and historians know of its cultivation throughout the Mediterranean region for several thousand years. Spanish settlers introduced it to California and Latin America. The word pomegranate derives from Latin and means "seeded apple." "*Granatum*" in French means grenade, a hand-thrown weapon.

Pomegranates have played an important role in the religious writings and paintings of Judaism, Christianity, and Islam. Both the Bible and the Qur'an mention the fruit with great respect and detail. The Hebrews used it as a symbol of fertility and abundance. In East Asia, it gets shattered at weddings to signify the same belief. Greek mythology attributes four months of winter to the four seeds that Persephone, daughter of Demeter the Goddess of Harvest, ate while kidnapped by Hades, the God of the Underworld, so she would be his wife.

© Kovaleva_Ka/Shutterstock.com

Pomegranates are peculiar, specialized berries because they do not have a fleshy fruit wall, yet they have numerous, fleshy seeds. The fruit is spherical, about the size of a large apple, with a thick, red shiny rind and a pronounced crown on top that represents the calyx region of the epigynous flower. Inside the leathery skin there are several partitions each with numerous shiny, pink or red pearly, juicy seeds. The partitions are yellow, dry, brittle, spongy, and bitter. The juicy outgrowths are from the aril of the seeds that are acidic and tart. The fruit rind is a good source of tannin, which the leather industry values. Pomegranates are eaten for the fleshy arils along with the seeds. Juice is most frequently consumed fresh, or made in to a syrup (grenadine syrup), wine, sorbets and jellies. Many parts of Europe and Asia have long used pomegranates in cooking. Pomegranate juice is a good source of vitamin C, folic acid, and antioxidant polyphenols. Due to its rich nutrient density and antioxidant benefits, the pomegranate is one of the **superfruits**.

Fruits That Are Drupes or Drupe-Like

Date fruits are from the date palm, *Phoenix dactylifera,* Arecaceae. Researchers believe that their native origin is the Persian Gulf. The cultivation of dates for their edible fruits has taken place since 4500 BCE in the Mediterranean region. The Arabs introduced dates to tropical parts of Asia, Africa, and Spain. Spaniards brought them to Central and North America. Iraq and Saudi Arabia are leaders in date cultivation.

Date palms are tall trees up to twenty-five meters tall with a crown of large, pinnate leaves that are about three to four meters long. Trees are dioecious; in order to obtain fruits a male tree must be in close vicinity. Modern agricultural trends use a machine to blow pollens on female trees. Fruits are drupes, oval in shape up to three inches long. The exocarp is thin and yellow to red in color; it turns papery brown and translucent when overripe. The endocarp is stony, and the edible mesocarp is brown, fibrous, sticky, and sweet. Many modern cultivars, including the seedless parthenocarpic varieties, get used for fruits.

Due to their being a source of nutrition and a beverage, the ancient desert-dwelling Bedouin people of Arabia considered date palms a "tree of life." Dates are rich in simple sugars and fiber and have a small amount of protein, calcium, and iron. Due to their high tannin content practitioners of folk medicine use dates as an infusion, either as a syrup or as a paste. Middle Easterners use dates in a multitude of ways; elsewhere people use them as a dried snack and in confectionery.

The **mango**, *Mangifera indica,* is one of several species of the genus *Mango* and the most well-known among the entire group. Anacardiaceae, a family that includes poison ivy, pistachios, and cashews, also includes mangoes. Mango trees are native to the forests of India, but people in adjacent tropics have widely cultivated them since ancient times. The Portuguese introduced the mango to Africa. Today people in the Pacific Islands, warmer parts of Australia, and the United States widely cultivate mangoes. The trees are medium to large, and the leaves are green, oblong, and coppery when young. Clear sticky sap from tender parts and leaves may cause allergic reactions and blisters on the skin. There are numerous varieties of mango that differ in fruit shape, sweetness of the flesh, fiber content, and so forth.

The fruits vary in size, ranging from round to oblong or oval with a pointed tip at one end. Mangoes are drupes; the endocarp surrounding the single seed is hard. Ripe mangoes have a yellow to red, thin leathery skin that one must peel with a knife. The mesocarp that tightly adheres to the skin is sour in green mangoes but ripens into orange-yellow, juicy, sweet-tangy, slightly fibrous flesh.

In southeast Asia, raw mangoes are cooked as vegetables to make sweet and sour chutneys and spiced pickles. You can eat mango flesh fresh or dried or make it into a puree, jam, and beverages, and add it to ice cream. In India, "**lassi**"—a beverage made from a mix of ripe mangoes, yogurt, and sugar—is a popular summer drink. **Amchur** is dried mango powder used in dishes for a sour taste. Hindus consider mangoes sacred, and their religious ceremonies start with mango leaves and fruits. They adorn doorways and prayer halls with garlands of mango leaves to signify the purity of the ceremony. Traditional South Asian medicine uses all parts of the mango plant from the sap to the mealy seeds to strengthen the nervous system and to treat rheumatism, anemia, and diarrhea. At times people eat roasted seeds and also use them medicinally to stop bleeding. Ripe mangoes contain antioxidants and are rich sources of sugar and dietary fiber, as well as vitamins A and C.

Rambuttan and **litchis** are drupe-like fruits. Their overall structure resembles a drupe even though both of them lack a hard stony endocarp characteristic of real drupes.

Rambuttan, *Nephelium lappaceum,* is a Malaysian fruit in the family Sapindaceae. It is native to Malaysia, and the trees are tall and spreading with compound leaves. Rambuttan is one of the most common fruits in Malaysia, but people in southeast Asia have cultivated it for several hundred years. The fruits hang in bunches, and each fruit is scarlet red in color, about two and a half inches long by two inches wide. The inedible outer rind is thin, pliable, and covered with soft, long, curved spines that give the fruit a distinct appearance. There is a single seed surrounded by fleshy, firm, sweet, translucent pulp (aril) with a mild flavor. You can enjoy the fruits fresh after removing the loose rind from the flesh, but you cannot eat the seeds and should discard them.

Litchi (or Lychee), *Litchi chinensis,* is native to South China, but growers have introduced it to adjacent Asian tropics and more recently into warmer parts of California, Florida, and Hawaii. The trees are round-topped, dense, and smaller than Rambuttan trees, with alternate, pinnately compound leaves. The fruits are round to oval, up to two inches long, and hang in small clusters. The inedible outer rind is thin and brittle. It is red at first, although it soon turns brown, and is slightly warty. Inside is a white, translucent, grayish or pinkish, sweet, mildly aromatic firm pulp (aril) surrounding a large, single, shiny brown seed, which one discards. American supermarkets and ethnic stores sell the canned aril and, occasionally, fresh litchis. Dried lychees that resemble raisins are also available. The Lychee is an important fruit in Chinese food traditions and the most popular fruit in China. Consumers consider lychee honey a high quality product. In moderate consumption, lychees relieve coughs and have a reputation as beneficial for ulcers, tumors, and gastralgia. Lychee- flavored ice creams and milk shakes are becoming popular in the United States.

Other Fruits

Pineapple, *Ananas comosus*, Bromeliaceae, is the second most popular tropical fruit. It is native to southern Brazil. Native people widely cultivated pineapples throughout Central and South America all the way to the West Indies even before Columbus arrived in the New World. Native Americans considered it a symbol of hospitality. The name derives from the fruit's resemblance to pinecones. Columbus introduced it to Spain, and later Portuguese, Dutch, and Spanish traders took it to the Philippines and other tropical regions. Warm parts of the world with good moisture grow pineapple successfully. J. D. Dole played a key role in encouraging pineapple plantations in Hawaii, the leading grower of pineapples in the world.

The plant is an herbaceous perennial with a short, stout stem that grows to a height of about four feet with a crown of leaves at the top. The strap-shaped leaves are thick and waxy with spiny margins. The inflorescence is a compact head with small red or purple-colored flowers. Each head has a few hundred individual flowers surrounded by a tuft of small, short leaves that makes up a crown. The pineapple is a multiple fruit composed of accessory tissue. A ripe pineapple is oval to cylindrical and about twelve inches long. The fruit includes the mature ovaries, flower stalks, perianth, and basal portions of the bracts that become fleshy and juicy. The rind is orange-yellow to reddish brown. When cut, it reveals a central fibrous core formed from the peduncle. Modern cultivars are parthenocarpic and lack viable seeds. Growers can propagate pineapples vegetatively by planting the leafy crown at the top of the pineapple, from underground portions of the plant, or from the suckers that originate at the axillary buds.

Pineapples are eaten fresh, also frequently canned, dried and used jam, juice, and candies. Pineapples contain a proteolytic enzyme, **bromelain**, found in commercial meat tenderizers. Native Americans made alcoholic beverages from the sweet juice and used it as a component of arrow poison and poultice. Pineapples are a good source of vitamin C and fiber. Folk medicine practitioners believe that pineapple is a diuretic and a treatment for inflammation, intestinal worms, arthritis, sore throat, and sinusitis, as well as for healing cuts or wounds.

Papaya, *Carica papya*, Caricaceae, is native to Central America, but now widely grown in all tropical parts of the world. Papaya is a fast growing, short-lived, large woody herbaceous tree that grows up to twelve feet tall. The huge leaves are deeply divided, each with a long petiole. Papaya plants are polygamous with male, female, and bisexual flowers often found on the same plant, sometimes at different seasons. The fruits fit into the pepo category except they derive from superior ovaries, unlike squashes and melons. Among commercial papayas two common types are distinguishable: The smaller type is Hawaiian, and the larger fruits are a Mexican type. Ripe papayas are mostly eaten fresh after peeling and removing the seeds. The orange yellow flesh has a musky flavor and soft smooth texture. One can also cook the firm, green ripe fruits as a vegetable. Other products include juice, sauce, puree, and jam. In the East Indies they cook the young papaya leaves as spinach greens. In India, they use dried black papaya seeds as an adulterant with black pepper. Papaya fruits are a rich source of iron, calcium, and vitamins A and C. One can tape the immature fruits for a milky latex that contains two proteolytic enzymes, **papain** and **chymopapain**. You can find papain as a main ingredient in commercial meat tenderizers, and it also has many practical applications in tanning, brewing, pharmaceutical, and cosmetic products. Chymopapain has medical uses, such as treating back pain related to pinched nerves or slipped discs.

Durian, *Durio zibethinus*, Bombacaceae, is the most famous Malaysian fruit that originated in Malaysia or in Borneo. It has been cultivated in tropical Asia for hundreds of years. It is a tall tree with relatively large

TABLE 4.7 Fruits that are becoming popular

Common Name	Binomial	Family	Description	Popular In
Malay apples	*Syzygium malaccense*	Myrtaceae	Oblong drupe, large cherry-like, reddish skin; flesh white, tart, crisp	Malaysia; Caribbean islands; West Indies
Mangosteen	*Garcinia mangostana*	Clusiaceae	Round, dark-purple, capped with calyx; rind thick; 2–5 seeds; aril edible, white, acidic and delicious	Southeast Asia; Jamaica
Mulberry —white, black	*Morus alba Morus nigra*	Moraceae	Multiple fruit, color varies from white, red, or black, sweet or tart	Western Asia; Europe; Eastern parts of North America
Noni/Cheese fruit	*Morinda citrifolia*	Rubiaceae	Multiple fruit with a pungent odor, oval about 4–5 inches long	Pacific islands; Southeast Asia
Rose apple	*Syzygium jambos*	Myrtaceae	Round drupe, cherry-like. Skin smooth, shiny, dull yellow to pink; flesh crisp.	East Indies, Malaysia, India; West Indies, Mexico, Peru
Sapota/Sapodilla	*Manilkara zapota*	Sapotaceae	Egg-shaped, brown, smooth-skinned berry. Flesh sweet, brown, and gritty	Mexico, West Indies; south Asia.

leaves up to nine inches long and three inches wide. The fruits are up to eight to twelve inches long and six to eight inches wide. They hang down from the leafless main or smaller branches on thick stalks. Outwardly it resembles a pineapple except the pericarp is thick and extremely tough and covered with sharp-pointed, coarse spikes. The fruit is a capsule with cream-colored soft, sweet flesh (aril) surrounding each large seed inside. The seeds, which are about three inches long and less than one-half inch wide, get discarded. Durian is well known for its peculiar smell from the aril that locals describe as delightful but that most foreigners consider offensive. People who have smelled it would describe it as a mix of garlic, onion, and rotten cheese or sweetened gasoline. The smell has very little connection to the truly rich, sumptuous taste of the edible flesh that only people who have tasted agree on. Malaysians hold this fruit in high esteem as is evident from a number of local names they give it, and they refer to durian as the "**King of all fruits.**" In Malaysia, the best fruits are eaten fresh, the remaining are used in ice cream, jam, and a local cake.

Fruit or Vegetable?

An official discussion of "fruit or vegetable" started in the late nineteenth century. The Tariff Act of 1883 in the United States required a 10% tax on all imported vegetables, not fruits. John Nix, a New Jersey based business-man, refused to pay the duty placed on his tomatoes that he imported from the West Indies. His argument was that tomatoes were only fruits and not vegetables. Anyone with a little botanical background might support his argument. But how does the law differentiate between a fruit and a vegetable? Let us see what the judge ordered when John Nix's case went to court in 1893. United States Supreme Court Justice Horace Gray, in his final decision, wrote that vegetables are usually served at dinner as a principal part of the meal, and people generally eat fruits as dessert. Come to think of it, many of us would not be excited to have tomatoes or okra for dessert, but we would be happy to have strawberries or honeydew instead.

Our daily vegetables such as cucumbers, beans, and squashes happen to derive from mature ovaries of the fertilized flower. Because they do not tend to become juicy and sweet, we cook them as vegetables. Or we use them as an accompaniment with our meals rather than eating them as dessert. Olives and avocado are two such fruits. We discuss some of these fruits most people around the world refer to as "vegetables" below.

Tomatoes, tomatillos, capsicum peppers, and **eggplants** are in the category of having the fruit type berry. They are in Solanaceae, a family that includes white potatoes and tobacco. Eggplants are native to India or southern China, whereas tomatoes and chili peppers originated in the New World and migrated to the Old World.

Tomato, *Solanum lycopersicum,* was first cultivated by the indigenous people of Mexico and was native to South America. The name tomato derived from the Mayan word ***tomatl,*** which later became tomato through the use of Spanish and later English. The wild forms of tomato were probably the cherry form. There are about five distinct species that contribute to a variety of tomatoes through cross breeding. Growers choose the modern cultivars for disease resistance and fruit quality, among other desirable qualities. Tomato plants are herbaceous annuals or perennials with a distinct offensive odor on the stem and leaves due to poisonous alkaloids, but the fruits are safe for consumption.

Spaniards introduced tomatoes to Spain in the sixteenth century, and the British introduced them to North America. The French and Italians were the first to accept them as a vegetable. Makers of Italian pasta, who depended on mostly cream-based sauces, soon invented a variety of recipes using tomato and chili peppers. By the middle of the 1500s, Italians had a fair use of tomatoes in their culinary practices. French called them love apples based on a word derived from an Italian name. The rest of Europe did not accept tomatoes readily, mainly due to their mistaken reputation as a poisonous fruit. European members of this family, such as deadly night-shade, were well known for their bitter fruits or hallucinogenic and toxic alkaloids. Germans rejected the tomato due to its repellent smell, calling it "wolf peach." The Philippines, Japan, China, and India became introduced to the tomato after Magellan's voyage in 1521. Tomatoes were popular in Asia before their spread in Europe, where they remained a garden plant restricted to glasshouses. Meanwhile, in the United States, French and Italian immigrants popularized the tomato. One of the events that changed the public's view about tomatoes seemed

Fig. 4.11 Chayote squash
© JIANG HONGYAN/Shutterstock.com

to have taken place around the 1820s in Salem, New Jersey when Colonel Robert Johnson announced that he would eat a basket of tomatoes in front of a crowd. The gathered crowd of about two thousand people became surprised and shocked to see the man who survived after eating tomatoes.

Tomato plants have travelled to space as an experiment to study the effects of space, radiation, and gravity on the germination and health of seeds. They were most commonly picked as one of vegetable plants for the International Space Station. Tomatoes are the first fruits to undergo genetic engineering, which resulted in a modified tomato, including the "Flavr savr" type.

Tomatoes play an important role in every cultural cuisine around the world. In addition to vitamin C, tomatoes are valued for the presence of a powerful antioxidant, **lycopene**, known to be beneficial for the prevention of prostate cancer and heart disease.

Tomatillo, a relatively recent introduction to North American vegetable markets, is a berry from *Physalis ixocarpa* that originated in Central Mexico. The fruits resemble a small, green tomato with a round shape. The persistent, inflated thin calyx surrounds and encloses the fruit. Fruits are commonly used in making sauces and in stews and extensively used in Mexican and Guetemalan cooking.

Capsicum peppers probably first became domesticated as hot peppers in Mexico. Botanists know of numerous varieties of *Capsicum annuum,* including the bell peppers commonly used as vegetables. Fruits are berries that are rich in vitamin C and dietary fiber with moderate amounts of protein. Chapter 8 discusses the uses of chili peppers as spices. Pimentos are sweet peppers that are juicy and flavorful; mostly used in canning and as a stuffing in green olives.

Eggplant, *Solanum melongena,* gets its name from the white, egg-shaped small fruits that were common several hundred years ago before the cultivation of the newer varieties that are purple. **Aubergine** is a European name for eggplant; in Africa and India it is **brinjal**. The fruits have a unique spongy texture and taste and tend to turn brown when bruised or cut. Early varieties were slightly bitter in taste, a reason for the slow spread of eggplant around the world. Recent varieties lack the bitterness and have come to take a leading place in European cuisine, including that of Italy, Greece, France, and Turkey. Even though eggplants are not a favorite in the United States, they are popular among Asian immigrants for one can see a variety on display in ethnic vegetable markets. The modern varieties come in different sizes ranging from small (Asian) to large (American). Eggplants contain a good amount of dietary fiber, folic acid, potassium, magnesium, and manganese.

Another family that provides us with a wide variety of vegetables is the squash family, **Cucurbitaceae**. It is hard to describe every single vegetable as there are too many kinds. Their common names are confusing and overlapping, as you can see in Table 4.8. You may frequently encounter many of these vegetables in local markets.

Olives, *Olea europaea*, **Oleaceae**, do not get cooked as a vegetable but rather used as an accompaniment to salads and as topping on pizza.

Native to the Mediterranean area, people have used olives since ancient times. The Bible, the Qu'ran, the Book of Mormon, Egyptian scrolls, and recorded literature in Greece and Rome mention them. Archaeological evidence suggests the use of olives in Italy five thousand years ago. People considered the olive tree sacred and used it as one of the symbols of the Greek goddess Athena. The Greeks considered the olive the symbol of goodness and nobility. We still use olive branches today as a symbol of peace. Olive cultivation started in California in the 1770s.

Olive trees are small, bushy, and about twelve meters tall; they have silvery green lance-shaped leaves. The fruit is a drupe, ovoid in shape, and about two inches long. Green olives are bitter due to a compound called **oleuropein**. Processing olives removes this compound to make them palatable. Drying, salting, or pickling

TABLE 4.8 Most common vegetables from the family Cucurbitaceae

Common name	Binomial	Notes	Origin
Bitter melon/ Balsam apple	*Momordica charantia*	Cucumber-like with warty skin; bitter flesh. Cooked and used in herbal medicine, especially for diabetes.	Uncertain; common in tropics
Calabash/Bottle gourd	*Lagenaria siceraria*	Vegetables; Dry rind is used in chicha cups, water vessels, ladles, horns, musical instruments, etc.	Africa; introduced to New World very early
Chayote/Chowchow	*Sechium edule*	About the size of an avocado, with deep ridges; single-seeded. Thin rind; flesh is greenish white, firm, bland-tasting. Eaten raw or cooked in stews and curries.	New world tropics; possibly Mexico
Cucumber	*Cucumis sativus*	Eaten raw in salads; pickled in vinegar and brine: whole or in slices	Asia
Luffa/Chinese okra	*Luffa acutangula*	Angular fruit, dry, brittle rind. Flesh is cooked in stews and curries: dries into natural sponge	Tropical Asia
Squashes—acorn, spaghetti, crookneck, some pumpkins	*Cucurbita pepo*	Mostly with a tough yellow rind; thick, firm, yellow-orange flesh, cooked in soups, pies, and stews.	New World tropics
Squashes—buttercup, turban, hubbard, winter	*Cucurbita argyrosperma*	Smaller kinds, eating and used for decorations; used in place of pumpkins	Uncertain; probably New World
Squashes—butternut, golden Cushaw, some pumpkins	*Cucurbita moschata*	Mostly with a hard yellow rind; thick, firm, yellow-orange flesh, cooked in soups, pies, and stews.	Uncertain; tropical Asia or New World tropics
Squashes—winter, winter marrow, some pumpkins	*Cucurbita maxima*	Mostly with a hard green to yellow rind; flesh smooth textured; used in pie fillings.	Uncertain; Asia or New World tropics
Wax gourd	*Benincasa cerifera*	Round, melon-shaped: thick rind is waxy. Flesh is white, and used in sweet pickles or curries.	Tropical Asia; Africa
Zucchini	*Cucurbita pepo*	Green or yellow; long or round. Always cooked.	New World tropics

Balsam apple

Calabash

Angled luffa

Slices of wax gourd

methods used in ancient times removed some of the bitterness. The first truly non–bitter, tender olives became processed in California in 1900 using sodium hydroxide (lye solution) that hydrolyzed the oleuropein and removed the bitter taste. True olive fans do not prefer these lye-cured olives as they seem to result in either spongy or hard, flavorless olives. A natural curing process involves a natural brine of salt, oil, and flavorings. For green olives, processers increase the salinity every two to three weeks by 2%. Black olives start from a lower salt concentration increased by 1–2% every two weeks. The taste of the final product depends upon the variety, the time of harvest, and the quality of the brine solution. Modern consumers are developing an appreciation for naturally cured olives. Such olives are available at deli counters or sold by specialized manufacturers.

Growers can harvest olives when immature if they desire green table olives. They may harvest the fruits left to ripen for processing or leave them longer if they desire oil. About 30% of the mesocarp contains oil,

Fig. 4.12 Fruits that are used as vegetables

Bell pepper: © Evgeni_S/Shutterstock.com; Bitter melon: © SOMMAI/Shutterstock.com; Olive: © Laboko/Shutterstock.com; Eggplant: © Svetlana Kuznetsova/Shutterstock.com; Pumpkin: © George Dolgikh/Shutterstock.com; Tomatillo: © bonchan/Shutterstock.com; Okra: © JIANG HONGYAN/Shutterstock.com; Tomato: © Tim UR/Shutterstock.com; Avocado: © Nataliia K/Shutterstock.com

mostly monounsaturated, which we know is beneficial for increasing the HDL level in the blood. Besides its use in salads and in pizza topping, olives are useful for extracting oil. Olives are also high in iron, vitamin E, and dietary fiber. Please see Chapter 7 for more information on olive oil.

Over the centuries growers around the world have developed many olive varieties. Some well known varieties: **Mission olives** are the common black olives, and Spanish olives, also known as **Manzanilla**, are the green olives stuffed with pimento, almonds, and onions. We know **Kalamata** olives as Greek olives, which are very popular in the United States because of their larger fruits, purplish brown color, and excellent taste. **Nicoise** olives are French olives, famous in a tuna salad recipe.

Avocado (Alligator pear), *Persea americana,* is in **Lauraceae**. Avocado trees are dense and fast growing. They are native to Mexico and Central and North America. Archaeological evidence suggests the use of avocado in southern Mexico since 7000 BCE. Aztec Indians called the avocado a "**testicle tree**" due either to the fruits that hang in pairs or to their stimulating properties. Avocado became introduced to California in the nineteenth century. Fallbrook, California leads the country in the cultivation of avocados and claims to be the "Avocado Capital of the World." Most tropical and subtropical countries grow avocados on a smaller scale.

Three main varieties of avocado are common in the United States: West Indian (large, thick, smooth skin); Guatemalan (large, thick warty skin); and Mexican (small, thin smooth skin). There are also several hybrids.

The fruits do not ripen on trees; one must pick them and allow them to ripen for about a day. The avocado is a one-seeded, specialized berry with a large, round, brown seed surrounded by a thin, brown endocarp. Some botanists consider this a drupe without a stony pit. Irrespective of its status in the fruit classification, the exocarp is normally thin, leathery, and inedible as well as either smooth and green or dark and warty. The edible mesocarp flesh is greenish-yellow, smooth, and creamy, with a light buttery taste. Ripe avocados never turn sweet; flesh is eaten fresh, sliced or chopped, in a sandwich, or in salads. Guacamole prepared from mashed avocado, garlic, lemon juice, tomatoes, and black pepper is a popular dip served in Mexican restaurants. Up to 30% of the fleshy pulp is oil, primarily **monounsaturated fat** known to promote heart health. The fruits are a rich source of fiber, potassium, phosphorous, magnesium, folic acid, and carotenoids, including lutein and vitamin C and moderate amounts of protein and iron.

Okra is *Ablemoschus esculentus*, **Malvaceae**, also known by the name "lady's finger." The exact place of origin is uncertain, possibly south central Asia. Ancient Egyptians used okra. The names okra and gumbo are African in origin. Okra is an annual plant and one of the highly heat-tolerant plants. Warm regions of the world cultivate it for its fruit that natives cook as a vegetable. People use okra in Middle Eastern, Portuguese, Spanish, Dutch, and Russian cooking. Travelers introduced it to the New World in the 1700s, and French immigrants later cultivated it. The French immigrants in the United States were extremely fond of cultivating it in Louisiana where they first settled. The fruit is normally green—although some are red-brown—angled, and about six inches long and tapered at one end. People normally pick the tender immature fruits for cooking and usually stew, steam, sautée, pickle, curry, or fry them. The slimy mucilage is a reason for okra's bad reputation. Acidity breaks the mucilage while cooking, so we recommend adding citrus, yogurt, or tomatoes in recipes. In India, people frequently use okra in cooking under the names **bhindi** and **vendi**. The mucilage acts as a natural laxative and helps maintain blood sugar levels; the fiber of okra is soluble and contributes to the health of the cardio-vascular and digestive systems. Okra is a low calorie vegetable and a good source of vitamins C and B6, calcium, and folic acid. Once mature, okra becomes a dry capsule and splits open. At that time, it becomes useless as a vegetable, but one can gather the seeds for further plantings and for seed oil that is rich in unsaturated fats.

Beans are a large group of vegetables derived from the family Fabaceae. Technically these are legumes picked for use as vegetables when they are tender. Fresh, tender pods or only the dry seeds get used in cooking. Members of this family have many other uses besides their use as vegetables. Chapter 5 will discuss them.

Fruits That Go "Nuts"

When you buy nuts from stores you are either buying the entire fruits or only the dry, edible seeds. The botanical meaning of a nut is different from what is on the labels. Chestnuts and macadamias are nuts; almonds and coconuts are not. Considerable discussion about this topic has occurred over the years. In fact, botanists still go "nuts" classifying the nuts. To bring some order to this confusion, let us describe a real nut. A nut is a single-seeded dry fruit with all three layers of the pericarp turning into one hard, stony shell. The hazelnut is one such nut. A number of the so-called nuts are actually the stony endocarps of drupes. This is true for almonds and coconuts where the fruit wall consists of separate layers of endocarp, mesocarp, and exocarp. Once the growers pick them from the trees, they remove the outer two layers, leaving the stony endocarp that protects the seed inside during packing and transport. There is also a drupaceous nut. The walnut, for example, has a leathery husk that eventually dries into a hard shell. Because of this, some botanists call walnuts nuts. Another argument is that it is a drupe. To satisfy both groups, let us refer to this and to a similar fruit, the pecan, as drupaceous nuts. In the case of the Brazil nut, the *nuts* are individual seeds of a large capsule with the seed coat becoming stony. Referring to the almond and to the coconut as nuts in everyday layman's language is acceptable but not quite appreciable in the botanical sense.

Fig. 4.13 Some of the "nuts"

Chestnut: © Iuri/Shutterstock.com; Coconut: © Sergey Kolodkin/Shutterstock.com; Pistachio: © Dionisvera/Shutterstock.com; Hazelnut: © oksana2010/Shutterstock.com; Pecan: © Dionisvera/Shutterstock.com; Almond: © Ljupco Smokovski/Shutterstock.com; Cashew nut: © Vinicius Tupinamba/Shutterstock.com; Walnut: © Luiscar74/Shutterstock.com

Real Nuts

The **macadamia nut** is from trees that originated in Australia. Its binomial is *Macadamia integrifolia,* a member of Proteaceae. Hawaii is the leading producer of macadamia nuts. The pericarp is extremely hard to break open, which partially contributes to the high price of this tasty nut. The edible portion of the nut is the seed.

The **cashew nut** is from small trees that are native to the northeastern parts of South America. Cultivation also takes place in Brazil and Peru. Its binomial is *Anacardium occidentale,* from Anacardiaceae, which also includes mango and pistachio. The Portuguese introduced cashews to India where they occupy an important role in local cuisine by the name of "**kaju.**" The edible seed sits inside a kidney-shaped, hard pericarp that contains a resinous sap that will cause blisters on the skin. Processers remove the seeds from the shell by heat processing that removes the resin. Another edible part of the cashew is the **cashew apple** that develops from the accessory tissue of the flower, the receptacle. Once ripe, it is obviously seedless (the seed is in the nut portion), yellow to red, and spongy with a sweet, musky flavor. The ripe fruits commonly sold in street markets in South and Central America, Asia, and Africa are safe to consume for most people.

The **hazelnut/filbert/cob nuts**, *Corylus avellana,* Corylaceae, are native to all temperate regions of the world. *Corylus avellana* is a widely cultivated European species. Leafy bracts with a fringed margin that resembles a beard surround the ripe nuts. This is the derivation of the name "fibert," which means "full beard." The nuts ripen around St. Philibert's day in Europe. Filberts are popular in most of Europe, especially in Italy and France, where they use the nut paste as a spread and the powder in ice cream and candies.

Chestnuts derive from at least three species of *Castanea,* a member of Fagaceae; all are native to the northern hemisphere. The recent cultivation in California is mainly from *C. sativa* (European) or *C. crenata* (Chinese). These are more resistant to the chestnut blight-causing fungus that wiped out the original American chestnut *Castanea dentata* plantations by the middle of the 1900s. Chestnuts are borne in a cluster of three, enclosed by fleshy, spiny bracts that spread apart once the nuts are fully ripe. Chestnuts are more popular in Europe where people eat them as a nut or a vegetable or use them in pastries and candies. Chestnuts and filberts are known from temperate regions of the world.

Drupaceous Nuts

Walnuts and **pecans** are in the walnut family, Juglandaceae. The trees of both are large, deciduous, and monoecious. As discussed at the beginning of this section, there is a controversy about whether walnuts and pecans are drupes or nuts. The reason, in part, is due to flower structure. In both trees, the flowers are small, inconspicuous, and surrounded by bracts and two smaller bracteoles that are partially fused with the calyx and inferior ovary. These fused structures later become a leathery **husk** that surrounds a **shell**—a mature ovary with a seed inside. The entire structure at this initial stage of development fits the description of a drupe—the shell being the endocarp and the husk being the outer two layers of pericarp. What happens later to this husk is the reason for the confusion: The husk dries into a hard, bony structure resembling a nut. This argument about "drupe or nut" might be of interest in a textbook of a different scope, but for our ethnobotanical curiosity, the fact that both are well represented in North America and used in making excellent desserts should suffice.

Walnuts are native to most temperate regions of the world and have wide representation. The commercial production of walnuts is from the Persian walnut, *Juglans regia. Juglans* in Latin means "Jupiter's acorn," and *regia* means "royal." It is native to southeastern Europe but widely planted in California. In parts of Europe, the people believed that walnuts warded off evil spirits because no other plants would grow in its close vicinity. Actually, walnut trees exhibit **allelopathy**, a chemical defense mechanism developed by certain plants. The seeds tend to become rancid because of their high oil content. Walnuts are rich in omega-3 fatty acids and valued for lowering LDL. Consumers use walnut husks to obtain a brown dye, and the plastic industry uses the seed oil in oil paintings. The Mediterranean dessert item baklava mainly uses walnut, even though almonds and pistachios are occasionally substituted in the recipe.

© Matthijs Wetterauw/Shutterstock.com

Pecans are in a different genus, *Carya*. Commercial pecans derive from *Carya illinoensis,* a native of Mexico and the American Southwest. Pecans are an important part of North American cuisine, largely due to the disease resistant and thin-shelled varieties. The mature husk is thick and leathery and eventually splits into four sections, exposing the inner shell. People eat the seeds fresh or use them in cooking. A traditional dessert in the United States item is pecan pie. Pecans are part of the recommendation of a healthy diet to improve the overall health of the cardiovascular system. They are a good source of vitamin E and proteins.

Nuts from the Endocarp of Drupe

Almonds are *Prunus dulcis,* a member of Rosaceae, subfamily Amygdaloideae, to which other stone fruits like plums, apricots, and cherries belong. Almonds are discussed in this section because most people know them as nuts. Almond trees are native to central Asia. The trees are small and deciduous with beautiful, large green foliage that tends to change colors as they age. Within the United States, they became introduced in the twentieth century. California grows most of our almonds and leads the world. Almonds are drupes, and commercial whole almonds are the hard pit of the fruit. Processers break the pits to remove the edible seeds that they then sell as "shelled almonds." Almond pits contain a cyanide precursor compound called "**amygdalin**" that provides the characteristic almond flavor and slight bitterness. High levels of amygdalin occur in other stone fruits like peaches where consumers do not eat the seeds. They eat the almond seeds raw or roasted or add them to cookies, in pastries, and in almond butter. A number of dessert items made from almonds include French macaroons, European nougats, and Indian badham kheer. Badham means almond, and kheer refers to any milk and nut-based sweet beverage. Ayurvedic and Unani medicines value almond oil. Almonds are rich in phytochemicals, monounsaturated fat, and vitamin E, and they contain a moderate amount of magnesium, phosphorous, zinc, calcium, and folic acid.

Pistachio nuts are the pits of another drupe from *Pistacia vera,* a member of Anacardiaceae. The trees are small, bushy, deciduous, and dioecious. They are native to the Persian region and most of central Asia, where people have cultivated them since 1000 BCE. The fruits are relatively small and hang in congested, heavy clusters. Iran leads in world production, followed by the United States. Many cultivars have been developed over the years. What the markets sell as nuts are the pits of the drupe that encloses a single green seed. The pits are beige in color, but growers used to dye them artificially red as a marketing strategy. This practice has stopped, and a natural dyeing process sometimes replaces it. People eat pistachios raw or roasted, and almost every major culinary practice around the world adds them to desserts. Pistachios are a rich source of monounsaturated fats, phytosterols, and dietary fiber, as well as a significant source of potassium, thiamine, and vitamin B_6.

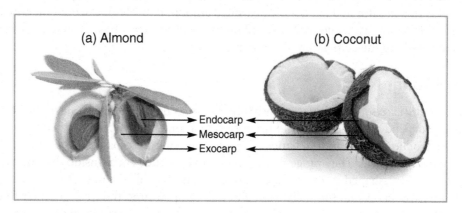

Fig. 4.14 Almond and Coconut

(a) © Zavodskov Anatoliy Nikolaevich/Shutterstock.com
(b) © homydesign/Shutterstock.com

The **coconut** is *Cocos nucifera*, a member of Arecaceae. The coconut palms are not real trees; as monocots, they lack the real woody tissue found in dicot trees. The smooth, unbranched trunk can reach a height of thirty meters, and the huge, long, pinnate leaves appear in a crown at the top of the trunk. A shorter tree variety called king coconuts is also common in Asian tropics where the people cultivate coconuts abundantly. Their native origin is uncertain even though fossil records suggest Indian origin. Coconut trees commonly grow along tropical beaches and inland. They have a highly branched and massive inflorescence with unisexual flowers.

© trindade51/Shutterstock.com

Coconut fruit is a drupe with a green or yellow, shiny exocarp, a fibrous mesocarp, and a hard, stony, brown endocarp that makes the pit. The coconut sold in markets is the pit with a single seed. Coconut seed is the largest seed known to humans so far. Unlike the nuts of dicots where the edible portion is the cotyledon in the seed, coconuts have an undeveloped cotyledon but a well developed endosperm that forms the "coconut meat." The edible endosperm occupies the hollow interior of the seed with distinct portions. One is the white, cellular endosperm that is the source of oil; the other is the liquid endosperm (coconut water). The coconut meat is grated and ground with warm water to make the "coconut milk" that is used in specialty recipes. Coconut water is slightly sweet, fat free, and best when the tender coconut is cut fresh. Depending on the size and maturity of the coconut, there may be up to twelve liquid ounces of coconut water. Young, immature coconuts have more water than older, mature ones. Processers crush the dried cellular endosperm called **copra** to extract coconut oil.

Every part of the coconut is useful to humans. We use the tree trunk for construction, the leaves for thatching, and the coconut meat and water for consumption. We use copra for oil extraction and fibers from the trunk and nuts to make shellfish nets, floor mats, baskets, and mattress stuffing. Coconut shells are useful in local craft items, and natives make a fermented beverage called **Toddy** or **Arak** by tapping the tree tops. Pressed oil cake is an excellent feed for livestock. Folk medicine practitioners use the astringent roots in intestinal ailments, the soft downy hairs from the leaves to stop bleeding, and the tree bark as an antiseptic.

Coconut meat is rich in saturated fat, fiber, and a small amount of protein. Coconut oil is loaded with saturated fat, and for this reason we do not recommend oil for cooking. The oil is favored in cosmetic industry. Indonesia leads the world in the cultivation of coconuts.

Other Exceptional Nuts

Brazil nuts, *Bertholletia excelsa*, Lecythidaceae, are from tall trees native to the forests of the South American tropics. The flowers depend on a special orchid bee for pollination. After successful pollination and fertilization, each tree can produce a few hundred fruits in a year. The pods are capsules about the size of a coconut pit, woody, heavy, and spherical. They range from three to five pounds each. Inside the pod there are about eight to twenty-four wedge-shaped stony seeds, each one mistakenly called a nut. The thick, woody pericarp opens up, leaving narrow cracks on the shell. Monkeys that jump around the trees get their fingers stuck in the cracks; hence the local name "monkey pods" to refer to these fruits. Merchants obtain Brazil nuts for trade entirely from the wild trees. The seeds are bland in taste and rich in fats with almost equal amounts of mono and polyunsaturated fats and a little less saturated fat. The amount of saturated fat is higher than that found in most nuts. There is a good amount of selenium and moderate amounts of protein, carbohydrate, magnesium,

© quentermanaus/Shutterstock.com

© zcw/Shutterstock.com

and thiamine. Brazil nut oil is useful as a lubricant, for making artists' paints, and in cosmetics. Many people also consider the timber of excellent quality.

Peanuts are not true nuts true but a special group of dry, dehiscent fruits called **legumes** that we discuss in Chapter 5.

Review Questions

Give brief answers in the given space.

1. Briefly explain how an accessory fruit develops. Cite two examples.

2. In what way did John Chapman contribute to North American fruit orchards?

3. Discuss how fruits played a role in the classification of Rosaceae.

4. What are superfruits? Cite a North American genus that gives us superfruits.

5. Explain the indigenous use of mango in its native origin.

6. Discuss how the banana travelled from the old world tropics to the new world tropics.

7. Why is blackberry an aggregate fruit and not a multiple fruit?

8. Explain the ethnobotanical significance of pomegranate, its origin and its indigenous and modern uses.

9. List at least three fruits from each of the following families that we use as vegetables: Cucurbitaceae, Solanaceae.

10. What is an aril? List at least three tropical fruits where the aril is edible.

Multiple Choice Questions

1. Botanically speaking, the strawberry is a(n) _____.
    A. berry
    B. multiple fruit
    C. aggregate fruit of many drupes
    D. combination of achenes and berries
    E. aggregate fruit of many achenes

2. Which South American fruit remained ill-famed in Europe and United States for a long time in history?
    A. Capsicum pepper
    B. Tomato
    C. Pumpkin
    D. Avocado
    E. Papaya

3. A true nut that also gives us an edible accessory fruit is the:
    A. almond
    B. walnut
    C. pistachio
    D. cashew
    E. coconut

Match the following fruit from Column 1 to the choices in Column 2. You may use a choice more than once or not at all.

Column 1

4. Guava

5. Mango

Column 2

A. Berry

B. Nut

C. Multiple fruit

D. Drupe

E. Aggregate fruit

6. In Hunza, a small kingdom in the Himalayas, _____ are an important part of the diet.
   A. apples
   B. coconuts
   C. strawberries
   D. apricots
   E. avocados

7. The source of oil in the olive is the:
   A. endocarp and the seed
   B. seed
   C. pericarp and seed
   D. mesocarp only
   E. leaves

8. "Amygdalin" is present in the pits of which of the following fruits?
   A. Cherry
   B. Peach
   C. Almond
   D. All of the above
   E. Only A and C

9. Even though people call these the "King of fruits," the smell and taste do not go together. Which one is it?
   A. Jackfruit
   B. Litchi
   C. Durian
   D. Papaya
   E. Passion fruit

10. The flower structure of which plant has a symbolic association with the crucifixion of Christ?
   A. *Citrus sinensis*
   B. *Mangifera indica*
   C. *Durio zibethinus*
   D. *Passiflora edulis*
   E. *Punica granatum*

True/false Questions

1. A cyanide precursor compound exists in all stone fruits of the rose family.   **True**   **False**

2. Citrus fruits are an important part of the American diet because their native origin is North America.   **True**   **False**

3. People believed walnuts had the ability to ward off evil spirits.   **True**   **False**

4. Chayote is a single-seeded squash.   **True**   **False**

5. Guava fruits have more vitamin C than most citrus fruits.   **True**   **False**

6. The avocado is rich in saturated fatty acids   **True**   **False**

7. Wild figs undergo a unique pollination mechanism.   **True**   **False**

8. Purple color and larger size distinguish Kalamato olives that are popular in the United States.   **True**   **False**

9. Brazil nuts are the endocarp of a capsule.   **True**   **False**

10. Most North American chestnuts are from a species native to North America.   **True**   **False**

CHAPTER 5
GRAINS AND LEGUMES

Most cultures around the world use grains and legumes as their main source of energy besides meat. Grains give the starch; legumes provide the protein. Both of them combined in a meal provide additional benefits in the form of fiber and vitamins. Two out of the "three sisters" of the Native American Indian meal include a grain (corn) and a legume (beans). In the first part of this chapter we will focus on some important grains widely consumed around the world as staple grains or as breakfast cereals. In the second half, we will learn about the legumes that provide us our vegetable beans, peas, and many other daily products.

Grains, also known as caryopsis, are dry, single-seeded, indehiscent fruits. Members of **Poaceae**, a monocot family, have grains as their fruits. The grain is actually the entire dry fruit and not just the seed. Corn kernel and wheat grain are all fruits with a single fused seed. Poaceae is the most important family that provides staple food not only for humans and also for their animals. **Cereal** is the edible grain of certain grasses. Cereal grains are the number one food source for humans; in addition, they give us a variety of products that are listed under each cereal grass. Fresh or dry plant stems and leaves are excellent forage for livestock.

In order to understand the importance of grains and to develop an appreciation for human intervention in the cultivation and selection processes involved in cereal grains, we must first learn grass plant structure and grain anatomy.

GRASSES

Grass Plant Structure

Grass plants are mostly herbaceous, with round stems, conspicuous nodes, and hollow internodes. Botanists call major branches of stems **culms**; they are usually hollow at the intermodal region. Leaves are alternate, usually two-ranked, and linear with a sheathing base that clasps around half of the internode. Grasses have a fibrous root system, and adventitious roots that arise from lower nodes also frequently support them. Perennial grasses have underground **stolon** or **rhizomes** that help the grass spread through vegetative propagation.

The floral features of grass plants are unique. Most grass flowers are bisexual, borne on branched inflorescences known as panicles. Each panicle further consists of a few to many **spikelets**, each with one to several **florets** (small flowers). Each floret basically consists of a reduced perianth called **lodicules**—three well developed stamens and a superior ovary with a single ovule and short style that ends in two long, conspicuous, feathery stigmas. Surrounding this entire incomplete floret are two bracts, an inner **palea** and an outer **lemma**. The tip of the lemma can extend into a long, sharp awn that helps in dispersal. Botanists also call the lemma

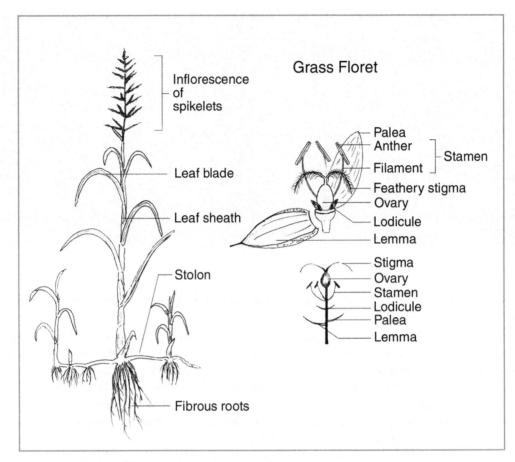

Fig. 5.1 Grass plant and floret structure
Courtesy Rani Vajravelu

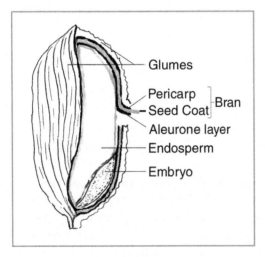

Fig. 5.2 Grain anatomy
Courtesy Rani Vajravelu

and palea by the names of bracts, glumes, or chaff. At the base of each spikelet, additional glumes are present and make the entire flower structure complicated.

Once the ovary has become fertilized, the fruit called grain forms. The pericarp and seed coat fuse together to form a structure called **bran**. Bran is rich in fiber and vitamins. The single seed inside has a small embryo called the **germ** and a well developed **endosperm** that makes up the bulk of the entire seed. The germ is rich in fats, proteins, and vitamins. The grass embryo consists of a single cotyledon (called **scutellum**) attached to an embryonic shoot with a coleoptile (forms shoot) and a coleorhiza (forms root). The endosperm is mostly starch, as in rice, sometimes with a little protein like in wheat. Inside of the bran and surrounding the endosperm is a layer called the **aleurone layer**. This layer is rich in protein and fats and aids the embryo during the germination process by breaking down the starchy endosperm into simple sugars.

The lemma and palea surrounding the florets now surround the mature grain. We call them glumes or chaff. Lodicules are inside of the chaff and remove easily when we get rid of the chaff.

From the Field to the Pantry

Once the grain is mature, farmers must harvest it. Historically, cereal grasses posed problems to farmers due to certain features unique to grass plants. The domestication and selective breeding of cereal grasses have considerably reduced these natural problems and have resulted in higher yields. For example, after heavy rains and wind, the tall hollow stems of grass plants tangled up, bent, and fell on the ground. This phenomenon called **lodging** would cause considerable loss because farmers could not easily harvest such lodged stems, even with machines. Modern cultivars of rice, wheat, and sorghum have shorter and stronger stems, thus eliminating this problem for farmers.

Shattering is a natural process in grass plants that allows dispersal of the mature fruit farther away from the parent plant. This posed problems in cultivated areas, resulting in loss of grains for the farmer. Modern selections of cereal grasses include the nonshattering varieties.

Since glumes are part of the floral structure, glumes or a hull surround a mature grain. One must hull these grains to free the edible grain for human consumption. It occurs by an activity called **threshing**, traditionally by hand or by applying external pressure on the harvested heaps of grains. To make this process easier for farmers, they selected free threshing cereal grain varieties for cultivation.

Winnowing is a process of removing the dry broken chaff mixed with the grains after the completion of threshing. Winnowing occurs by holding baskets of the chaff-grain mixture over the head and slowly pouring it down to separate the heavier grains from the lighter fragments that blow away. These form a heap in a distant place, depending on the wind force and direction. Obviously, it is a time-consuming job that takes patience and strength. Modern machines called **combines** thresh and winnow the harvested grains in the field. In developing countries, small farmers still use the hand winnowing method.

Polishing the whole grain is relatively easy in today's world of machines. But in ancient times, people did it by hand by placing the hulled whole grains in between two stones or in a wooden pestle and mortar type device to create constant abrasion by rolling and beating to scratch off the bran. Later, workers would blow off the fine dust of bran and germ in small batches, separating the polished grains. All this hard work took place in farms or in every household. They considered the bran and germ mixture, however nutritious, as poor people's food or as feed for bulls and cows. They preferred the polished grains as food for humans. In our health-conscious modern society, people pay extra to buy bran and germ from the stores.

A **whole grain** is everything in a grain except the chaff and lodicules. It includes bran, the endosperm, and the embryo, thus making it a healthy staple. A **polished grain** is when bran, the aleurone layer, and the embryo get removed, resulting in a white refined grain. We also know this process of polishing as **pearling**. White grains and their products are less nutritious, but they have a longer shelf life due to the removal of fat contained in the embryo and in the aleurone layer.

A great deal of discussion is going on currently about the health benefits of whole grains in daily diet. Until recently, those determining the authenticity of commercial whole grain products such as bread and cereal largely based their conclusions on the product labels. Consumers had many unanswered questions about the actual amount and quality of whole grain they were getting out of these labeled products. The Whole Grains Council, a nonprofit consumer advocacy group, helps consumers find whole grain products by placing their whole grain stamp on the package.

Fig. 5.3 Whole grain stamps

Courtesy of Oldways Preservation Trust and the Whole Grains Council, www.wholegrainscouncil.org

Rice

Rice as a staple grain feeds billions of people worldwide, more than other grain. The people of Asia have eaten rice the longest in history. Rice is a symbol of fertility, and several places in Asia also consider it sacred. Even people in Western civilization are familiar with the tradition of throwing rice at weddings. Rice belongs to a genus *Oryza* that includes several distinct species, out of which, *Oryza sativa* is globally more important. Western Africa has cultivated *Oryza glaberrima* since 1500 BCE. The plants are native to southeastern parts of Asia. Rice plants are about one and a half meters tall and have a less complicated spikelet structure compared to other cereal grasses.

Rice cultivation started in the Middle East around 300 BCE and spread to northeastern parts of Africa. Rice was unknown in the Middle East and Europe until Alexander the Great conquered some lands in the east. It took nearly fourteen centuries for rice cultivation to start in Spain and Italy. The Portuguese introduced rice to Brazil, then rice cultivation began in Western Africa. Britain started to import rice from the islands of Madagascar by the sixteenth century. The introduction of rice into the United States was accidental. In 1695 a British shipment of rice landed in a port in South Carolina due to an unplanned change in the route. Even though rice cultivation began soon after, it took nearly three centuries to start commercial cultivation. With the introduction of nonshattering rice varieties the yield increased six times within forty years. California, Arkansas, Louisiana, Texas, and Mississippi are major producers of rice in the United States. Besides China and India and many Asian countries, Brazil is a major rice producer. Latin American populations also largely depend on rice as an important food item.

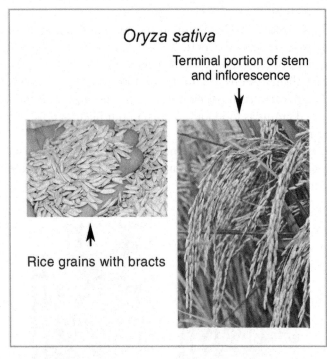

Oryza sativa

Terminal portion of stem and inflorescence

Rice grains with bracts

Fig. 5.4 Rice plant with grains
© Worraket/Shutterstock.com
© FrameAngel/Shutterstock.com

There are three major types of rice, each one favored for a particular type of preparation. The *indica* type is the long-grain rice of the United States, preferred for its non-sticky grains that remain separate after cooking. The *japonica* or *sativa* type is the short-grain rice that sticks together when cooked; and, finally, the *javanica* type yields a larger grain rice but is not as well-known as the other two types.

Cultivation methods involved in rice vary based on the areas that grow it. The **paddy** grows rice in standing water. **Upland rice** uses the slopes or flat fields of raised lands with only a sufficient amount of water needed for growth.

Rice endosperm is mostly starch with a cellulosic wall that helps the grain to retain its shape even after cooking in boiling water. Rice by itself is bland with a mild starchy fragrance that makes it ideal for use in combination with other flavorful items to create a variety of food items ranging from spiced rice pilaf and broth to fried snacks. Eating polished rice alone as a main source of energy for long periods of time without the addition of vitamin rich vegetables leads to **beriberi**, a health condition that results in nerve inflammation.

The rice bran is a natural source of vitamin B1 (thiamine), which the polishing process removes. Even though eating enriched polished rice or parboiled rice can overcome this problem to a certain extent, the best alternative would be including unpolished, **brown rice**. Recent research has led to the development of genetically modified **golden rice** that has a pigmented endosperm. The golden color is a result of the beta-carotene, a precursor for vitamin A.

Rice is one of the gluten-free grains. Besides cooking the entire grains for a fluffy starch source, one can make the uncooked rice into flour that has a variety of uses, including rice noodles. You can add cooked rice itself to vegetables, meat, spices, or milk to create a variety of savory or dessert items. Rice bran is a source of oil considered healthy for cooking.

© Oksana Shufrych/Shutterstock.com

Parboiled rice gets steamed along with the glumes that enable the starch to absorb the nutrients. Once dried with the glumes removed, some nutrients remain in the grain even after polishing. Parboiled rice takes longer to cook, does not stick, and retains a dull color.

Basmati rice is a fragrant, long grain rice preferred for its long, distinct cooked grains for making meat or vegetable-based Middle Eastern and Indian spiced rice dishes called **biriyani**. **Jasmine rice** is the Thailand fragrant rice with a nutty aroma; it has nothing to do with jasmine. **Red rice** is a fermented, red-purple colored rice used in place of regular white rice or as food coloring; in pickled items and in Chinese medicine. It acquires its red color through the inoculation of red yeast *Monascus purpureus* into the cooked rice, which then gets dried or packed in jars as wet paste.

The United States' Rice Federation that serves as the global advocate for all segments of the U.S. rice industry lists at least eight different rice varieties grown within the United States. Some of them are U.S. Aromatic rice, Arborio rice, sweet rice, and sushi rice.

Wheat

Wheat is an ancient grain mentioned in the Bible as the grain used by Adam and Eve to make their first bread. All of our modern wheat varieties originated from a single genus *Triticum*. Botanists believe the wild species is *Triticum monococcum*, which is native to Iraq and Syria. Wheat cultivation has spread around the world, and China is leading in the cultivation of wheat. The Spanish brought the wheat to the New World in the 1520s, and English colonists tried it in North America without much success. After the 1870s wheat cultivation proved successful when Russian immigrants started cultivating it in Oklahoma and parts of Texas. Extensive research and studies in biotechnology have helped to improve the pest resistance and nutritional aspect of wheat. Modern milling techniques have also improved the quality and quantity of wheat products, such as the flour, germ, and bran, which separate easily and sell for various uses.

Original wild forms of wheat were diploid (2n = 14) like most plants. Natural hybridization and mutations followed by human selection resulted in several new beneficial varieties. Some of the varieties became incorporated into cultivated forms for a better yield, for ease of harvesting, and for the ability to make raised bread. For example, **einkorn** (one grain) wheat is the modified form of an early mutant that showed a suppressed shattering. **Emmer** wheat, *Triticum turgidum* var. *dicoccum*, is a tetraploid that resulted in a free threshing natural

Wheat plant
© Zeljko Radojko/Shutterstock.com

mutant type known as **durum** wheat, *Triticum turgidum* var. *durum*. Hexaploid wheat that arose from a cross between emmer wheat and another uncertain diploid species is the **bread** wheat, *Triticum aestivum*. As the number of chromosomes increased, so did the amount of beneficial proteins, called **gluten**, that help the wheat dough rise when mixed with yeast and result in a fluffy, raised bread.

Farmers grow different varieties specifically and use them for different purposes. The most extensively grown types are durum and bread wheat. Bread wheat is the dominant bread variety grown in the world used in breads, breakfast cereals, and pastries. Durum wheat grows in the northern United States, parts of Europe, and in India. People prefer it for making noodles, pasta, and Indian flat breads, known as chapatti.

Wheat flour is an important ingredient in European and North American cuisine in bread and pastry. It also plays a major role in other cultural cuisines as well, ranging from Italian pasta to Indian flat breads. For this reason, we call wheat "**the staff of life.**" Flour is a major ingredient in most food items. There are various types of wheat flour depending upon the parts of the grain used in making the flour and other components. **Whole grain flour** is ground from the entire grain using the bran, germ, and the endosperm. This flour is rich in fiber, protein, and starch. **Germ grain flour** comes from the endosperm and the germ, leaving out the bran. This flour is similar to the above except for the lack of fiber. **White flour** contains only starch because it comes only from the endosperm, leaving out the fiber rich bran and protein rich germ. **All purpose flour** is a blend of different wheat flours with a moderate amount of protein. **Enriched flour** has vitamin B1, iron, and niacin added to plain white flour. **Graham flour** is a blend of finely ground endosperm mixed with coarsely ground germ and bran flours. **Self-rising flour** is wheat flour (white or whole wheat) pre-mixed with chemical agents that help in self rising when the cook makes dough or batter. **Semolina flour** comes from durum wheat.

Corn

Corn, *Zea mays*, is a New World native crop that grows well in both temperate and tropical regions. The basis of Mayan, Incan, and Aztec civilizations was corn. Due to its close association with native Indian life, this plant has the name "Indian maize." *Zea* in Greek means "to live." Archaeological evidence suggests that corn initially became domesticated in Mexico. By the time Europeans arrived in the New World in the 1500s, corn was a staple food for the Native people. Without a formal education in agronomy native Indians were well aware of the mutual benefits of growing beans along with corn. Bean plants helped enrich the soil with nitrogen due to their symbiotic association with *Rhizobium* bacteria in root nodules. In addition, eating meals comprised of beans and corn offered health benefits by bringing important nutrients together. Corn was introduced to Europe through Christopher Columbus. It did not gain popularity in Europe even during the Irish potato famine in the 1840s. It spread to other parts of the world from there, and people still commonly cultivate it.

Corn plants have unisexual flowers unlike other cereal grasses that have bisexual flowers. The plants bear male flowers in terminal branched inflorescences and female flowers in the axils of leaves. We call each female

Fig. 5.5 Sand paintings, an art form developed by the Native Americans of the Southwest, often portrayed the four plants sacred to them. Clockwise from the upper left: corn, squash, beans, and tobacco.

inflorescence an **ear**; it has several rows of female flowers on a fleshy central core (together called a **cob**) surrounded by several tightly packed bracts leaves called **husks**. Each female flower has a long, silky style known as **silk** that protrudes out of the husk.

Growers choose modern corn cultivars for various uses including flour, popcorn, and eating types. Corn is gluten free and not as nutritious as wheat. It is deficient in amino acids tryptophan and lysine, and is relatively low in protein. **Pellagra** is a disease characterized by dementia, diarrhea, and dermatitis caused by tryptophan and niacin deficiency. People in North America, parts of Africa, and Europe showed these symptoms when their diets consisted mostly of corn without the addition of missing nutrients.

Native Indians used corn husks to make braided baskets, mats, and other craft items, including cornhusk dolls and dried cobs as fuel. Native people ate the corn silks and folk medicine practitioners used them in folk medicine. Modern society depends on corn as much as native Indians did several centuries ago. One can eat corn kernels (fruits) raw or cooked before they are fully ripe and mature kernels cooked whole or popped as popcorn. One can also dry and make them into corn flour. **Corn flour** has a mild flavor and use as a thickening agent, as starch, and in baking. **Cornmeal** is coarse whole grain corn flour. People frequently use it in tamales in Mexico and in tortillas in the United States. The biofuel ethanol and biodegradable plastics are made from corn. Corn is used in baby food and cereals. When roasted, the kernel becomes a coffee substitute. Corn syrup is a sweetening agent used in baking, candy making, and carbonated beverages. Corn pollen is useful in specialty cooking. Corn oil is extracted from the germ of the grain. The remaining residue is used

Fig. 5.6 Corn plant
(a) © Anton Foltin/Shutterstock.com
(b) © Maks Narodenko/Shutterstock.com

as animal feed. Beer, whiskey, cake mixes, snack foods, toothpaste, glue, shoe polish, ink, marshmallow, cosmetics, paint, varnish, paper products, and pharmaceutical products such as antibiotics and aspirin are some products that use corn.

Oats

Oats, *Avena sativa,* are native to Eurasia. Major grain crop fields have grown oats as secondary crops. Their domestication started about 1000 BCE, probably in Europe, which is relatively recent when compared to other cereal crops. People used oats primarily as feed for horses and considered them unsuitable for humans. Romans cultivated oats mainly for their animals and considered oats as poor people's food. Oats now grow throughout the cooler parts of the temperate regions that receive a good rainfall. You can distinguish oats from other cereal grasses by the open spreading inflorescences with large pendulous spikelets.

Oat grass stems make excellent livestock feed; farmers feed oats to cattle, chickens, and, of course, horses. Dried oat stems called straw are soft and absorbent, one of the items prized in animal farms as bedding. Today, 95% of all oats grown here in the United States get used as animal feed.

Oats have a hard outer hull formed by the glumes that one must remove before consumption. Hulled oats, called **groats**, have their bran intact, and you can eat them unrefined. At this stage, oats can be crushed or rolled to make a flat, flaky grain. Sometimes the grains are cut in to two or three pieces before rolling. We call these quick rolled or **steel cut** rolled oats. Since we can eat oats unrefined, they make an excellent source of protein, fat, and soluble fiber. Oats contain an excellent blend of amino acids with a good amount of lysine, which is absent in many cereal grains. Oats have a relatively high amount of soluble and insoluble fiber content that is valuable in lowering cholesterol. They also have a high amount of polyunsaturated fatty acids and vitamins B and E, and they provide more calories than wheat.

Oatmeal is rolled oats used in making porridge, cookies, and bread. Oats have very little gluten, and you cannot use them in bread making directly. The Scottish and Irish use oats in their daily cooking; especially in Scotland every celebration uses an oat-based dish. Oat flour can be used in muffins, breads, beverages and desserts. In the United States oats are an important part of a healthy breakfast, and they have found a place in many grain-based cereals. Skin care products, cosmetics products, and skin lotions use powdered oat talc.

Sorghum plant
© suthiphong yina/Shutterstock.com

Sorghum

Sorghum, though not well known in the United States, is popular in other parts of the world. Its binomial is *Sorghum bicolor*. It is native to Africa and was probably in cultivation in Ethiopia before 3000 BCE. From there, it spread throughout Africa and to India. It was grown in U.S. since 1800s, but its cultivation increased after World War II. People in the United States commonly refer to it as **sorgo**, in Africa as **great millet**, and in India as **cholam**. The plant resembles corn morphologically except for bisexual flowers borne on terminal inflorescences. The plants are extremely drought tolerant. Sorghum is the fourth most important cereal grain besides wheat, rice, and corn. There are three main uses for this plant: syrup used in making molasses; the stem and leaves as forage for cattle; and grain used in a very similar manner to wheat in India and Africa, but not in the United States. Botanists generally recognize three or more varieties. One is for the sweet syrup or for forage; another is for making brooms from extensively branched inflorescences; and another is mainly grown for the grains. Due to the absence of gluten, the flour is not suitable for making baked breads.

In India and Africa, they make flat breads from the grain flour and use the grains in making beer, butyl alcohol, and **silage** (food for cattle). They use the sorghum grain starch in adhesives, for gums in stamps and envelopes and as sizing material in paper and textiles.

The sorghum variety with large, juicy stems contains up to 10% sucrose. Farmers harvest such juicy stems and crush them to extract the syrup that they later boil and distill to obtain the thick, brown, sticky sweet syrup called **molasses**. One can also manufacture sugar from sorghum but not at the same level as sugarcane. When grown for making brooms, the fruiting inflorescence of the broom-corn cultivar variety is preferable because of its long, straight side branches. They are cut and the grains removed. The remaining stalks are gathered and made into **whisk brooms**. The "**Broom Corn Capital of the World**," which makes these brooms, is Arcola, Illinois. In India and elsewhere, they use the grains as popcorn and also as animal feed. Consumers use sorghum widely in beer making, especially in Africa.

Other Edible Grains

A number of grains get used in making other commercial products including the barley used in beer making. Table 5.1 lists some of the grains and their major uses.

TABLE 5.1 List of additional grains

Common Name	Binomial	Origin	Major Uses	Notes
Barley	*Hordeum vulgare*	Southwestern Asia	Baking, Beer, Whiskey making	One of the oldest domesticated crops
Millet–Finger	*Eleusine corocana*	Africa	Flat bread, porridge, in beer	High in protein and amino acids
Millet–Pearl	*Pennisetum glaucum*	Africa	Flat bread, cooked, in beer	Drought resistant
Rye	*Secale cereale*	Southwestern Asia	Pumpernickel bread, rich source of lysine. Rye whiskey, Dutch gin; forage	Secondary crop Poor person's wheat
Triticale	*X Triticosecale*	Artificial hybrid	Bread; replaces Rye in most recipes	Combines desirable characteristics of Wheat and Rye
Wild rice	*Zizania palustris*	New World	Cooking	Expensive due to demand

Sugarcane

Sugarcane is not a grain-yielding grass, but we include it here because it is a member of Poaceae and similar to Sorghum. Its syrup-filled stems give us the common table sugar. There are about six perennial species of the genus *Saccharum,* out of which two species still occur in the wild today: *S. robustum* in New Guinea and adjacent Pacific Islands and *S. spontaneum,* which has a wider distribution. It ranges from northeastern Africa to the Pacific and through Asia with its greatest concentration in India. The other four species occur only in cultivated forms. One of them is *S. officinarum,* also called **noble cane**, which originated on the islands of the

Finger millet
© Matjoe/Shutterstock.com

Pearl millet
© Tukaram Karve/Shutterstock.com

south Pacific, most probably New Guinea. Most of the modern sugarcanes are cultivars of natural hybrids from the above species of *Saccharum,* each with several clones of their own.

Brazil is the major cultivator of sugar cane. Sugarcane fields and sugar processing refineries exist around the world today. The methods of growing sugarcane and processing sugar were technologies transferred first from India to China in the seventh century. According to historical documents, two sugar makers were sent to China along with the Buddhist monks based in The Mahabodhi temple of Bihar, India.

India was probably the first country that used the sugarcane to make sugar. The process of making unrefined sugar has existed since about 3000 BCE. When Alexander the Great visited India around 350 BCE, he was surprised to find Indians making honey (sweet syrup) out of the reeds (grasses). People of the Hindu religion still celebrate an ancient tradition as a way of showing their respect to the sun god. This is a harvest festival celebrated throughout India under different names, but all of them coincide with the harvest of grains. In south India, where the festival is a four-day celebration called **pongal** (based on a Tamil word meaning "to boil"), they cook rice in decorated clay pots outdoors under the sun surrounded by sugarcane and other sacred plants, such as turmeric and mango leaves. The highlight is letting the rice and water mixture boil over the pot to signify abundance and prosperity. The tradition calls for raw sugar from sugarcane fields and freshly harvested rice to make a sweet dish, also called pongal. Even though modern festivals substitute store bought ingredients, the celebrations continue with the same spirit.

Flower honey was the principal sweetener in Europe until after the 1500s. Even though sugar was a desirable commodity in Europe, people could not cultivate it due to unsuitable climatic conditions. The scarcity and popularity of cane sugar among Europeans led to the high price of imported sugar. It was very expensive and cost almost about US $110 for a pound of sugar. People confined sugar use to making medicines more palatable.

Columbus brought sugarcanes to the New World in the late 1500s and initially introduced them to the present-day island of Haiti and The Dominican Republic. Soon after, the entire West Indies started to grow sugarcane successfully. Sugar became an important American export, thus reducing its scarcity and lowering prices. By the end of the twentieth century, the price of sugar had come down drastically (to one pound of sugar for about US$1.40), mainly due to increased production.

By the eighteenth century triangular trades had become successfully established between the West Indies and other countries. In the famous American sugar triangle, New England in the United States and West Africa became involved. The countries of the West Indies were shipping raw sugar to New England refineries.

Fig. 5.7 Sugarcane
© phloen/Shutterstock.com
© pittaya/Shutterstock.com

Merchants sent the rum made from raw sugar in the sugar refineries of New England to Africa to buy African slaves, whom they brought to the West Indies to work in the sugar fields. Historians call this the famous **American sugar triangle**. By the 1700s approximately sixteen thousand Africans had become enslaved and transported to work in the sugarcane fields. The British sugar triangle involved England instead of New England as the northern point. Instead of rum, they sent small pieces of jewelry, ornaments, beads, and cloth to Africa to buy slaves.

The British Parliament imposed a duty for sugar in its **Sugar Act** passed on April 5, 1764. This encouraged protests among the colonists who were already going through a time of economic depression right after the seven year French and Indian war. This also encouraged smuggled sugar among the colonists in Connecticut. When strict enforcement patroling the shipments and customs occurred, the colonists burned and sank one of the ships in the harbor. This event preceded the Boston Tea Party of 1773.

Farmers propagate sugarcane from cuttings rather than from the seeds. Modern methods of vegetative propagation start by means of **sett**. A sett is part of the lower stem with a few lateral buds at the node and tissues for adventitious roots. When you plant a sett, the lateral buds sprout into new stems. The adventitious roots initially support the plant, which later gives out tillers that grow into new stalks. You can harvest stems every year, but each successive harvest gives a smaller yield, and eventually new setts must replace it for a better yield.

Sugarcane plants have smooth skinned, thick, juicy, fibrous stems that come in various colors ranging from deep purple to green, striped, and so forth. The stems are about ten feet tall with a crown of thin, linear leaves. Growers first **top** the juicy stems to remove the leaves. Alternately, they can burn the leaves at the same time to evaporate the water content. They clean the cut stems and crush them through rollers to express the juice. They then boil the juice, clarify it with lime (calcium hydroxide), filter and evaporate it, and finally place it in a centrifuge. After this, they separate the crystals called raw sugar and the syrupy molasses. **Raw sugar** still may have impurities and is brown and sticky. This gets used to make molasses or rum or is sent to refineries to make the final refined product, **white sugar**. The refined white sugar involves many steps. The crystals get re-dissolved into heavy syrup and centrifuged to remove the impurities. Next, the processers add diatomaceous earth to clarify the syrup and once again filter and send it for crystallization. When crystallized sugar gets spun in hot, revolving drums, it breaks into **granulated sugar**. Finely ground crystals yield **powdered sugar**. **Confectioner's sugar** is powdered sugar with added cornstarch to prevent clumping. You can form **sugar cubes** by pressing the slightly moistened sugar into cube molds. **Brown sugar** is refined sugar with molasses gets re-added to give the brown color.

Besides white sugar and related commercial products, we can use sugarcane leaves and leftover crushed stems as animal feed and fuel. In Brazil, ethanol extracted from sugarcane provides fuel for automobiles. Scientists are seeking other sources of biofuel besides sugarcane and corn. One other promising plant is the American native plant called switchgrass, a species of Pennisetum that is still being explored for this possibility.

Bamboo is another member of Poaceae valued for its wood products. Chapter 11 discusses it because of this reason.

Quinoa plant
© Paul S. Wolf/Shutterstock.com

Amaranth plant
© mimohe/Shutterstock.com

TABLE 5.2 Grains that are not "real" grains

Common Name	Binomial and Family	Nutrition	Fruit Type and Notes
Amaranth	*Amaranthus spp. including A. caudatus;* Amaranthaceae	Rich in protein; contains lysine, iron; no gluten	Achene; ancient food source for Aztecs
Buckwheat	*Fagopyrum esculentum;* Polygonaceae	Contains quality proteins; fiber; no gluten	Achene; popular in Middle East
Quinoa	*Chenopodium quinoa;* Amaranthaceae	Rich in protein; lysine, manganese; moderate amounts of magnesium	Achene; ancient food source for Incas

Grains That Are Not "Real" Grains

Not all the grains labeled as "grains" are from the grass family. The fruits of such "grains" may resemble a grass grain to an untrained person. Many of these are actually a single-seeded achene. Table 5.2 gives a list of most well-known "grains."

LEGUMES

Legumes are dry, dehiscent fruit pods of a dicotyledon family, Fabaceae. The traditional name for this family is **Leguminosae**, based on the legume fruit found in included members. Fabaceae is a large plant family further subdivided into **Faboideae**, **Caesalpinioideae**, and **Mimosoideae** based on the differences in flower structure. A large number of useful products like timber, gum, dyes, oil, medicine, fiber, fruits, and vegetables derive from all three groups. Of these, Faboideae provides most of our vegetable protein contained in plants' pods and seeds. This portion of the chapter will discuss such sources. Please note the word **legume** may denote either the plants or the fruit pods, depending on the context.

Faboideae includes herbs, shrubs, climbers, and trees. The distinguishing features are the composition of the leaves in pinnate or trifoliate fashion and flowers that are butterfly-shaped and zygomorphic. The roots associate with nitrogen-fixing bacteria like *Rhizobium* that live in the **nodules** and simplify the atmospheric nitrogen into usable forms for the legume plant. This symbiotic association is mutual as both partners benefit from each other. Because of nitrogen-fixing bacteria in their root nodules, major crop fields grow legumes as alternative crops.

Pulses and Lentils

This portion of the chapter focuses on the fruit pods (legumes). People use many legumes for their green, young edible pods with immature seeds. For example, one can cook string beans, pole beans, and snow peas along with their seeds and pericarp. In some, the pericarp is thick or leathery and unsuitable for cooking. Many such pods, when mature, contain seeds that one can either cook immediately or dry and store for later use. We refer to such dried edible seeds of cultivated legumes as **pulses**. Pulses are next to cereals in their importance as human food. The cultivation of the pulse crop is an ancient practice in both the Old and New Worlds. For people who practice vegetarianism, pulses are the main source of protein. The seeds have a low moisture content and dry seed coat to keep them in storage for a long time.

Lentils are pulses derived from several varieties of a legume plant, *Lens culinaris,* which is a native of the Near East. The seeds are usually sold as lens-shaped individual pieces, each one is a cotyledon. Lentils are sold whole or split, with or without the seed coat. All of these form the basis for such local names as black, green, red, and yellow lentils. It is common to refer to seeds of other genera as lentils too, and the seller and the consumer do not consider it a serious offense as long as they are aware which lentil is desirable in a recipe. Lentils are a rich source of protein and used extensively in making some type of soup or saucy dish in every single cuisine, especially around the Mediterranean region. In India, where the majority of people practice vegetarianism, the use of lentils in everyday meals is a common practice. They give the name **dhal** to all of these lentils in general. Also, dhal refers to curried, saucy dishes made primarily out of lentils.

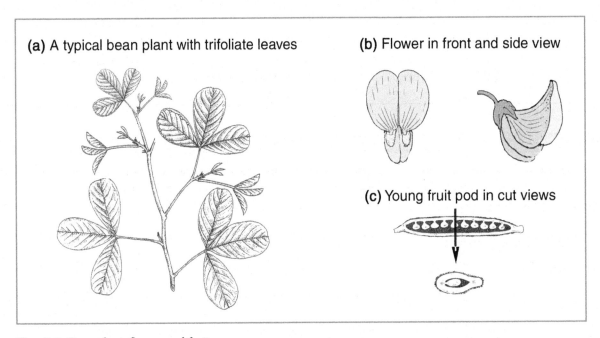

(a) A typical bean plant with trifoliate leaves

(b) Flower in front and side view

(c) Young fruit pod in cut views

Fig. 5.8 Bean plant, flower, and fruit
Courtesy Rani Vajravelu

Beans and Peas

What we call peas and beans are legumes. Their use as whole pods or only as dry seeds is common in every cuisine. There is a huge variety of peas and beans; no single distinguishing feature will segregate them easily. Common names are confusing and sometimes overlap. Table 5.3 summarizes some of the well-known peas and beans.

Some legumes provide additional products that are of commercial value. We describe a few such legumes below.

Peanut

Peanut, *Arachis hypogaea,* is also known as **ground nut** and **goober**. This plant is native to the Peruvian region and widely cultivated throughout South America. Archaeological excavations of coastal regions of Peru dated 800 BCE revealed some ancient crafts and jewelry with a peanut motif. By the middle of the fifteenth century, peanuts were common in the West Indies. The Portuguese introduced them to West Africa where their popularity spread around with the name "goober." Africans popularized peanuts in North America. Spaniards introduced peanuts to the Old World and took them to the Philippines. They spread from there throughout China, Japan, Malaysia, and India. Tropical and subtropical regions of the world now grow peanuts widely. China is the major grower of peanuts today. Consumers use the seeds as a snack food and as a source of oil. After extraction they use the remaining residue, called peanut cake, as food for cattle.

Peanut plants are annual herbs, and their stems may spread or trail. The leaves are distinct in two opposite pairs. The flowers are yellow with an ovary stalk (**gynophore**) that elongates after fertilization and grows towards gravity, eventually burying the developing fruit under the soil. The rest of the maturation process occurs under the soil, which takes about five months. The pods develop prominent ridges and a brown color due to a lack of sunlight. Each pod has two to six seeds. Mature pods and gynophores remain buried underground along with roots until pulled out during harvest time. The structure and development of the peanut is the reason for its common name of groundnut and the basis for its species name *hypogaea*.

TABLE 5.3 Commonly used Peas and Beans

Type	Common Name	Binomial	Uses	Origin
Peas	Black-eyed peas/ Cowpeas	*Vigna unguiculata*	Used in good luck dish on New Year's Day in southern U.S.	Africa
	Chick-peas/Garbanzo	*Cicer arietinum*	As high-quality protein; flour; infant formula	Northeastern Africa
	Peas, green, sweet, snow, sugar snap	*Pisum sativum*	Entire pods or only the seeds used in cooking	Near East
	Pigeon peas	*Cajanus cajan*	In United States split pea soup; Indian dhal dishes	India
Beans	Broad beans	*Vicia faba*	As vegetable; may cause favism (due to anemia) in some people	Mediterranean
	Common bean: string, kidney, pinto, navy, green, snap, wax	*Phaseolus vulgaris*	All kinds of vegetable beans, widely used around the world	Probably South and Central America
	Lima beans	*Phaseolus lunatus*	Seeds as vegetable	South America

Cicer arietinum   *Cajanus cajan*

Fig. 5.9 Chick peas and Pigeon pea plants
© Sever180/Shutterstock.com
© Tukaram Karve/Shutterstock.com

Fig. 5.10 Peanut plant
© Amawasri Pakdara/Shutterstock.com

In the United States, many southern states grow extensive peanut crops, including Oklahoma, which boasts an extensive cultivation of peanuts by placing a peanut statue in the city of Durant. The popularity of peanuts in the United States is in part due to the enterprising efforts of an agricultural chemist named George Washington Carver (1864–1943), who discovered three hundred uses for peanuts and introduced crop rotation using peanuts in cotton fields.

Less than half of the peanuts in the United States are used in making candies and snacks. The rest are used in making sweetened peanut butter, widely used not only in the United States but also in Canada, South and Central America, Australia, Africa, The Netherlands, and Saudi Arabia. South Americans made the first ground peanut butter for local consumption. A St. Louis physician (believed to be John H. Kellogg in the 1890s) prepared a ground paste somewhat similar to the modern product for his sickly patients who had difficulty chewing normal food. Modern commercial preparations started in 1922 when J. J. Rosefield developed a smooth creamy paste marketed widely under the names Peter Pan and Skippy. In southeast Asia, they make a salted or spicy version of ground peanuts as a temporary paste to use in cooking.

Botanists classify several cultivars based on the growth habit, number, and oil content of the seeds. The two most basic ones are the **Spanish type** with higher oil content and the **Virginia type** with larger seeds and lower oil content that consumers roast and use in snack items. Roasted or boiled groundnuts are eaten as snacks. They are rich in protein and vitamins B and E. Chapter 7 describes peanut oil.

Soybean

Soybean, *Glycine max* is a twining or prostrate plant. Modern cultivars are erect with trifoliate leaves and white to purplish flowers. Soybeans are one of the world's most important food sources now, but they did not gain popularity outside of China and Japan for a long time in history. The Chinese held soybeans in high esteem. They probably originated and got domesticated in northeastern parts of China around 1000 BCE People in eastern Asia used soybeans initially as an important food crop. They ate the seeds as a vegetable and used them in making soy milk and in infant food and soy sauce. Miso, tempeh, and tofu were other fermented products of the soybean that are making their way to the Western world through the popularity of Asian cuisine.

© Fotokostic/Shutterstock.com

Soybeans were introduced to the Western world in the middle of the eighteenth century, but they did not gain attention. Their use in the United States started only after the 1920s when the heated plants proved safe for animal consumption. Since then the cultivation of soybean has increased dramatically, and the United States is the number one supplier of the soybean and its products. Because of this sudden popularity, people refer to the soybean as the **"Cinderella crop."** Henry Ford in the 1930s was one of the Americans, besides George Washington Carver, who promoted the soybean's use in paints, plastics, fabrics, and in food. The cuisine in the United States has not openly accepted the soybean as a vegetable. People cannot consume soybeans raw due to the presence of the enzyme trypsin that interferes with the digestion of other proteins. Cooking or blanching denatures trypsin and makes the beans palatable. More than half of the United States' production of soybeans gets exported to other countries; the rest gets processed into oil and other commercial products listed in Chapter 7.

Soybeans come loaded with beneficial phytochemicals that have profound health effects on the human body. These include the phytoestrogens, known to alleviate menopausal symptoms in women. They play an important role in promoting heart health, bone density, and cancer prevention. Besides high amounts of protein and essential amino acids, the soybean is packed with calcium, iron, zinc, phosphorus, magnesium, B vitamins, fiber, and omega-3 fatty acids. It also gives rise to a number of other commercial products. Like the cereal grain corn, the soybean is the source of many household products and surpasses every other legume in its usefulness to humans. **"Cow without bones"** is a nickname given to this most versatile plant that provides everything from meat to milk. Soy protein is a high quality protein almost similar to the proteins found in meat products. Because of this, manufacturers use soy protein to make texturized vegetable protein (TVP) used in imitation meat, such as burger, hot dogs, and bologna.

The following two legumes are members of Caesalpinioideae, which yields edible pods. Many members yield valuable timber for use in the wood industry.

Tamarind

Tamarindus indica is a tall, spreading tree with long, brown, pods that hang in clusters. People in southeast Asia and Africa knew of the trees for a long time, and wild plants are common in both areas. Tamarinds got introduced to the Middle East from India. The name derives from "tamar hindi," which translates as **"Indian date."** The legumes differ from other legumes in having a fleshy, brown, tart, and sometimes sweet, sticky pulp

Fig. 5.11 Tamarind plant and mature pods

© joloei/Shutterstock.com
© Sivaaun/Shutterstock.com

surrounding the seeds. The dry pods have a brittle exocarp that one must remove manually to reveal the pulp. The dried pulp can get stored for long periods of time due to its high acid content. The pulp is used in making sour-tasting curries, chutneys, sweet and sour sauces, and beverages, as well as in pickles that almost every south Indian household commonly makes. Latin American countries use tamarind in snack items and in beverages. The seeds are brown skinned and hard, and one can boil or roast them and eat them as a snack item.

Carob

A tree similar to tamarind is **carob**, *Ceratonia siliqua,* which is native to the Mediterranean region and known for its pulpy legume. The mesocarp is sweet, and one can eat it raw. In ancient times, people used carob seeds as units of weight for small quantities of precious gemstones because they were extremely uniform in size. The word "carat," our present-day international unit for the weight of diamonds, derives from the name "carob" in reference to carob seeds. One carat is precisely 200 milligrams. Roasted seeds are used in a beverage. The dried mesocarp pulp is ground to make carob flour which is used as a chocolate substitute in candy, ice cream, brownies, and cookies. Locust bean gum, a thickening agent in ice creams and salad dressings, comes from the ground seeds of the carob tree.

Forage Legumes

A number of legume plants get used as important **forage** plants. Traditionally, farmers have grown them along with the forage grasses to complement each other. Several hundred forage legumes exist around the world. Some grown in the United States are alfalfa, *Medicago sativa*, Clover, *Trifolium* spp., and *Melilotus* spp. Out of these, people have cultivated alfalfa since ancient times as a forage crop and grow it extensively around the entire world. Because of this reason, they refer to alfalfa as the **king of the forage legumes**.

Review Questions

Give brief answers in the given space.

1. Describe the unique structure of a typical grass floret.

2. Explain how human selections have improved the cultivation of cereal grains.

3. Imagine having brown rice for dinner today. Reflect on the processes that went on after its harvest in the field until it reached your plate.

4. You have probably heard that whole grains are better than polished grains. What is the scientific reason to support this statement?

5. What is durum wheat? What are its main uses?

6. Which cereal grain is one of the "three sisters" for Native Americans? List at least five current uses of this grain.

7. How is sugarcane incorporated in the traditions of India?

8. Compare and differentiate pulses and lentils.

9. Explain the processes in groundnut formation that make its name appropriate.

10. State any one legume with a pulpy mesocarp and at least two of its uses.

Multiple Choice Questions

For each question, choose one best answer.

1. The "bran" of grains consists of _____.
   A. bracts
   B. lodicules
   C. seed coat and fruit wall
   D. bract and embryo
   E. endosperm only

2. A "culm" is _____.
   A. a solid stem of a dicot
   B. nodal region of a monocot
   C. hollow internodes of a monocot
   D. hollow stem of a dicot
   E. bract in a monocot flower

3. When you are eating white bread made from refined flour, what part of the grain are you mostly consuming?
   A. Embryo
   B. Endosperm
   C. Aleurone layer
   D. Chaff
   E. Whole grain

4. "Corn silk" is what part of the corn plant?
   A. Entire female flowers
   B. Staminate flowers
   C. Bracts of the corn ear
   D. Style part of female flowers
   E. Chaff and lodicules

5. Beriberi, a nerve inflammation, is common among people who eat mostly _____ as a staple food without the addition of other nutrients in their diet.

   A. whole wheat
   B. white rice
   C. corn
   D. refined wheat flour
   E. unrefined oat

Match the following crops that are native to the given choices A through E. You may use a choice once, more than once, or not at all.

Column 1

6. Peanut

7. Sorghum

Column 2

   A. Pacific islands
   B. South America
   C. Himalayan foothills
   D. Africa
   E. Ireland

8. This particular state in the United States, _____, boasts of its extensive production of peanuts by placing a peanut statue in _____.

   A. Missouri; St. Louis
   B. Oklahoma; Durant
   C. Ohio; Cleveland
   D. Georgia; Atlanta
   E. Texas; Houston

9. A St. Louis physician first prepared "peanut butter," a favorite of American kids. Which of the following reasons was the motive?

   A. To provide nutritional supplement to starved children in America
   B. For expectant mothers as a source of protein
   C. For patients who had difficulty chewing regular food
   D. For patients who suffered from stomach disorders
   E. To supplement a vegetarian diet that was low in fiber

10. Kidney beans and string beans are in the genus _____.

   A. *Pisum*
   B. *Phaseolus*
   C. *Vigna*
   D. *Lens*
   E. *Vicia*

True/false Questions

1. Fabaceae is a large family further subdivided based on the fruit type.   **True**   **False**

2. Wheat is "food for billions."   **True**   **False**

3. "Paddy" may refer to rice raised in wetlands with standing water.   **True**   **False**

4. The activity of removing the grains from the bracts is lodging.   **True**   **False**

5. "Wild rice" belongs to the same genus as the long grain rice.   **True**   **False**

6. Triticale is the hybrid between wheat and rye.   **True**   **False**

7. Dhal is an Indian name that may refer to the dry seeds of a legume.   **True**   **False**

8. Golden rice is a genetically modified rice.   **True**   **False**

9. Oats were secondary crops meaning they were second in nutrition compared to wheat.   **True**   **False**

10. Sugarcane plants use sett for vegetative propagation.   **True**   **False**

CHAPTER 6
PLANT-DERIVED BEVERAGES

Some plants are naturally equipped with special chemical compounds in the leaves, fruits, or seeds. When we ferment or brew such plant parts in water following simple recipes, the result is a variety of refreshing beverages. Such plant-derived beverages are an integral part of any cuisine whether traditional or modern. Packed with an array of flavors and health benefits, they make interesting substitutes to plain water. Around the world, people consume plant beverages alone or along with a meal to help wake up in the morning, for refreshment during the day, or to wind down after a long day. Human civilization has evolved around plant beverages; consequently, these plants have spread from east to west and vice versa. They have played a role in developing friendship and hospitality from the earliest times and continue to do so even in today's busy lifestyle. Historically, they have helped to establish religious practices; to improve social skills; to enhance intellectual powers; and, in some cases, to change the financial status of the country. Considering the wide variety of beverages from plants, we have divided this chapter into two main sections: non-alcoholic, naturally caffeinated beverages that include the most widely known teas to the less known mate and guarana; and alcoholic beverages that include well known wines to the least known chicha.

Caffeinated Beverages

Tea

Tea is the second most popular beverage in the world besides water. Tea is prepared from the young processed leaves of a short shrubby tree, *Camellia sinensis,* a member of Theaceae. Tea plants are native to the foothills of the Himalayas in the area adjoining Tibet, India, China, and Myanmar and cultivated elsewhere in tropical and subtropical areas of the world. We know about the discovery of tea as a beverage from various legends from China and India, including the story of a Chinese emperor, Shen Nung, whose boiled water was accidentally brewed with the dried leaf of a nearby bush. An Indian legend describes Bodhidharma, an Indian prince who chewed on the leaves of the tea plant to stay awake during his seven years of continuous meditation and prayer. Another story is about the origin of tea from the eyelids of a pious Buddhist monk.

Brief History

Historical documents from the time of Confucius prove that the Chinese knew tea as a beverage before 500 BCE. Tea was the national drink of China by the seventh century CE. It was customary to send the finest tea bricks (at that time they pounded and pressed tea leaves into a brick-shaped mold) to the emperor's court as a tribute.

A Chinese scholar, Lo Yu, wrote a detailed account in the eighth century of tea and extensive accounts on preparation methods in his treatise Ch'a Ching (The Tea Classic). It was most popular in China and also much read in Japan. Japanese Buddhist monks probably introduced tea to Japan in the eighth century. In the twelfth century, a Zen Buddhist monk named Eisai introduced the tea ceremony. He later wrote the first Japanese book on tea, the Kitcha-Yojoki (Book of Tea Sanitation). The method of preparing tea by stewing cured leaves began in the Chinese Ming Dynasty around the fourteenth century CE and became popular among Europeans and in the Western world. The processing methods for different kinds of teas now known as black tea, green tea, and oolong tea also developed under the Ming Dynasty.

Even though the Buddhist monks in China and Japan initially revered tea as a stimulant for aiding in meditation practices, the habit slowly started to spread outside of the religious circle. Observation of the tea ceremony became an expression of social elegance and sophistication. The early ceremonies in Japan unfortunately mixed gambling and alcoholic beverages. Thanks to the influence of tea masters, especially Sen No Rikyu in the sixteenth century, who refined and formalized the rituals of tea ceremony that is still being followed to this day.

The traditional English tea party has some resemblance to this old tradition with the hostess sharing an elegant setup and sophisticated chit-chat with the guests centered on the best homemade tea meticulously prepared.

Tea reached Britain in the seventeenth century through Dutch traders and soon became the beverage of wealthy and aristocratic people. London coffeehouses played an important role in taking tea to the common public that was getting accustomed to this new stimulating beverage. The British East India Company established in western India supplied much of the tea, and the export increased from a mere one hundred pounds in the 1680s to fourteen million pounds by the 1780s. The colonists who arrived from the Netherlands and England in the 1650s introduced tea to North America. With its high cost of about $30 to $50 per pound, it was obviously the beverage of fortunate wealthy people initially. Despite the high cost, by the middle of the 1700s, colonial Americans had quickly adapted to the taste of tea. They used special ceramic and silver teapots and cups, and most households brewed tea.

In order to avoid the high prices and duty imposed on tea by the British East India Company, the colonists preferred buying tea for cheaper prices directly from smugglers. This smuggling trade had a huge impact on the profits of the East India Company. To combat the problem, the British Parliament passed a **Tea Act** in 1773 that allowed the British East India Company to monopolize the trade by selling tea directly to the colonists at a lower price. This also curtailed the local business by smugglers. The colonists were against the new Tea Act and showed their protest by boarding the ships in Boston harbor and dumping the precious cargo of tea. In the early morning of December 16, 1773, some men disguised as Mohawk Indians boarded the four ships, one after another, opened the containers, and threw the tea into the sea. This incident, called the **Boston Tea Party**, contributed to a series of events that united the colonists and eventually led to the American Revolutionary War and the Declaration of Independence signed in 1776.

India, Sri Lanka, China, Japan, and Taiwan are the five major regions of the world known for the cultivation and export of a wide variety of teas. The Himalayan foothills of northeastern India produce the best quality tea.

Variations in Tea

In ancient China they brewed tea with onion, ginger, orange, and crumbled rice. Since then, brewing has undergone enormous changes. Modern society prepares tea one cup at a time. Consumers prefer either loose leaf fragments or tea bags for this quick alternative to old-time tea kettles. Whatever the method, everyone involved favors the finest tea, favoring plantations with higher elevations and abundant rainfall for the best tea leaves. The plants are pruned and kept at an optimum height to encourage hand picking of the tip leaves done on clear mornings to keep the leaves tender yet fresh. Such harvested tip leaves go through a processing system that will yield white, black, green, or oolong tea. Pekoe, Orange Pekoe, and Flowery Pekoe are three grades that indicate the fineness of the tea leaves.

White tea is the tip leaves of certain cultivars fast dried to retain their silvery white color. The beverage prepared from white tea is actually pale yellow. The Chinese have known about white tea for thousands of years, but the rest of the world was introduced to this beverage recently. It has less caffeine than even green tea and contains more nutrients because of minimal processing.

Black tea leaves are oxidized and yield a dark beverage with a stronger flavor and more caffeine. In order to obtain black tea, the growers spread the harvested tip leaves in racks where hot, dry air passes for ten to twenty hours. This process, called withering, removes the excess moisture from the leaves. After this, a machine rolls the leaves, releasing the enzymes to activate the fermentation process. The rolled leaves are spread out in cool fermentation rooms for several hours until they turn a coppery color. Another drying process called firing takes place by passing the leaves through hot air chambers, resulting in black, dry leaves.

Green tea leaves undergo a minimal oxidation process and yield a green beverage. Green tea contains less caffeine than black tea and more polyphenols with antioxidant properties. Growers obtain the green tea leaves by steaming the harvested tea leaves then rolling and firing. Since the leaves are not fermented, they remain green. The Chinese have known about green tea for more than twenty-five hundred years. Much of the green tea comes from China, India, and Sri Lanka. Extracts of green tea are available in capsule form; the dosage and effects are uncertain.

Oolong tea is prepared from semi-oxidized (or semi fermented) greenish brown leaves. The name oolong derives from the Chinese meaning "black dragon tea." It was popular in China during the Qing Dynasty (that followed the Ming Dynasty in the fourteenth century). The preparation of oolong tea requires great skill and time and results in a large spectrum of tastes. There are several variations in oolong teas, each with its unique characteristic and aftertaste.

Brassica tea is the latest invention (2004) in tea, resulting from research at Johns Hopkins University in Baltimore. They added an extract of SGS (sulfuraphane glucosinolate) in broccoli (a member of Brassicaceae) to either black or green tea. Scientists hope that this tea will combine the beneficial effects of broccoli and tea.

Iced tea is brewed tea poured over a glass of ice, a popular American drink. Richard Blechynden first introduced it to the American public at the St. Louis World Fair in 1904. He was trying to sell Indian and Sri Lankan black teas on a hot summer day. As a marketing strategy, he poured brewed black tea into a glass of ice and served it to a thirsty crowd on a hot, summer day. It was a hit right away and has remained so until now.

A **tea bag** is a convenience device that serves as a tea infuser in most households of the Western world. A New York merchant introduced tea bags in 1904 by wrapping loose tea in hand-sewn silk bags. William Hermanson of Boston invented heat-sealed tea bags in 1930. Modern-day disposable tea bags are made from plant based fibers, one main source being the abaca fiber derived from banana leaf fibers. Broken leaf pieces of tea fill the bags.

Fig. 6.1 Camellia plant with flower
© Elena Mirage/Shutterstock.com

Instant tea in the form of a dry paste with a mix of tea leaves, sugar, and milk was developed in 1885 by John Brown. The more recent dry powder kind invented in the 1940s has achieved popularity in the United States.

The Flavor of Tea

Whatever the type of tea, consumers prefer tea for its flavor. The tea flavor is due to an essential oil, theol present in the tea leaves. The stimulating effects are due to alkaloids, theophylline, and caffeine. Theophylline has positive effects on asthma and respiratory symptoms. Compounds such as tannins that cause the brown stain on the teapot and on the teeth of excessive tea drinkers also exist in tea. Tannins are responsible for the dark color in black tea.

Flavored tea is a popular modern twist to the traditional tea flavor even though the Chinese knew of scented teas with flower petals by the thirteenth century. Thanks to modern technology, with the addition of a flavor ranging from jasmine to pomegranate, one can create any imaginable flavor in a tea. **Chai** is a generic name for tea mostly used in India, Iran, and some other parts of the world. People in India prepare a flavored variation of chai (masala tea) by slowly brewing the tea with a mix of aromatic spices, milk, and sugar. The commercial version of chai is a readymade powder or a concentrate that is much different from the original preparation, which calls for freshly grated ginger and cardamom, and, if desired, cinnamon and cloves.

Tea Facts

- ◆ China is the leading producer of tea followed by India.
- ◆ Consumers refer to Darjeeling tea as the "Champagne of Tea."
- ◆ Tea arrived in Europe in 1590 through a Portuguese Jesuit missionary.
- ◆ Tea contains half the amount of caffeine found in coffee.
- ◆ Tea is an excellent source of antioxidants that help in cancer prevention.
- ◆ Experts recommend up to four cups of tea a day for optimum health benefits.
- ◆ Consumers in the UK prepare up to 96% of their tea from tea bags.

Coffee

Coffee is one of the most valuable and popular beverages worldwide. The discovery of coffee was more recent than tea. Coffee plants are native to Ethiopia but now grow widely in the tropics of Africa, India, Central America, Colombia, Brazil, Java, and Malaysia. The earliest known records of use are from the eighth century in Ethiopia, where goat herders chewed the leaves and berries to keep hunger and fatigue away. Once people discovered their stimulating properties, they ground coffee beans and mixed them with fat to use as a staple. Coffee, as a brewed beverage, comes from the "beans" (two seeds) of fleshy coffee berries known as "cherries." A very small percentage of coffee berries end up having a single bean inside the fruit. Such a single bean is a **pea berry**. They are more rounded and roast differently, and workers must separate them from the rest of the **flat berries** during processing.

Brief History

Coffee arrived in Yemen before the fourteenth century and rapidly spread to Egypt, Italy, and England by the mid-seventeenth century. By 1675 there were about three thousand coffee houses in England that functioned as meeting places for the exchange of news, business, and politics for people, all for a penny a cup. Customers dubbed these coffeehouses "penny universities," and one such place later became Lloyd's of London.

Arabs monopolized the coffee trade until the Dutch managed to secure seeds from the borders of the Red Sea and started plantations in the East Indies. They sent trees from this plantation to the botanical gardens of

Amsterdam and later on to Paris. There is mixed information about whether the plantations in Suriname, Haiti, Antilles, and French Guiana started independently through the Dutch and French or through a single plant transplanted to the Caribbean island of Martinique by a French infantry captain. Whatever the predecessor, Brazil's plantations started from borrowed seeds of coffee from one of the above places.

Coffee Plants

Coffee plants are evergreen, perennial, small bushes from Rubiaceae. They prefer abundant rainfall and cooler subtropical and tropical climates. The

Fig. 6.2 Coffee plants with flowers
© ntdanai/Shutterstock.com

small, white, fragrant blossoms are bisexual. The red, shiny, two-seeded berries derive from inferior ovaries. Once ripe, growers pick the berries, remove the seeds, and dry and roast them. The processing and roasting methods result in a variety of commercial grades. There are several species of coffee, but *Coffea arabica* is the most prominent species of coffee, widely used and predominantly grown in Latin America. *C. canephora* (or *C. robusta*) yields about 25% of the coffee in the world. It is known for its higher caffeine content than *C. arabica*. Grown in Africa and Asia, consumers use it mostly in decaffeinated and instant coffee. *C. liberica* is known for bitter coffee, cultivated in African lands usually at lower altitudes. Leaf rust (*Hemileia vastatrix*) is a fungal pathogen that attacks coffee plants that limits coffee plantations in certain areas.

Coffee Processing

The **wet processing** method yields superior or mild coffee. Growers wash and pulp ripe coffee berries by pressing the fruits through a screen. Pulping removes the exocarp and most of the mesocarp. They remove the remaining amount of pulp clinging to the seeds by allowing it to ferment and then washing it off with a large amount of water. Alternately, they can also accomplish this process with mechanical scrubbing. Fermentation in coffee is an enzymatic, chemical alteration of several compounds that leads to the fine flavor of coffee when roasted. After the scrubbing process, the growers spread the beans evenly and dry them either under the sun for about a week or by using a machine to avoid mildew formation. They then remove the crumbly thin dry parchment (endocarp) by a hulling process using a machine.

The **dry processing** method yields natural or hard coffee. Most of Brazil's coffee uses the dry method. Growers spread out the ripe coffee berries thinly in the sun and turn them frequently to ensure even drying and to prevent mildew formation. This takes about fifteen to twenty-five days. They store the dried coffee berries in bulk and send them to the mill to remove all outer layers (hulling), including the thin endocarp, in one single step.

Producers grade coffee beans based on the country of origin, the region, and the altitude, the processing methods, and the bean quality, including the size and density. The International Coffee Organization (ICO) is responsible for setting and regulating the production and price of coffee for its member nations. Among other responsibilities related to international coffee issues, it brings together coffee producing and consuming countries through international cooperation to improve the standards of living in developing countries.

Roasting

The roasting method determines the fine aroma and flavor of coffee. The growers dump the dry, clean, green coffee beans into horizontal, hot revolving drums called roasters where the beans heat anywhere from a few minutes to up to thirty minutes. They can control the temperature to be between 370°–482°F. A newer method called a fluidized bed roasts coffee on a cushion of hot air. Mild, sweet coffee results from lower temperatures whereas strong, bitter coffee derives from higher temperatures. During the roasting process, the starch in the seed endosperm begins to convert to sugar. As the temperature increases, the sugar caramelizes and then carbonizes, resulting in such dark, black beans as black French roasts. The color of coffee beans changes as well from yellow to brown or to dark brown.

Caffeol, an essential oil, is responsible for the coffee aroma produced by roasting coffee. Caffeine is a stimulant alkaloid concentrated in coffee beans. The caffeine content is independent of the strength, flavor, and aroma of the coffee. Coffee leaves also contain a small amount of caffeine.

Variations in Coffee

Decaffeination is the process of removing the alkaloid caffeine from the unroasted coffee beans. German chemist Ludwig Roselius invented the first commercial decaffeinated coffee in 1903. He used salt water and benzene to remove the caffeine. Concerns about the use of benzene made this process unpopular. An alternative to this method was the European process that uses methylene chloride to absorb the caffeine from the beans, but health concerns led to its ban also. Swiss water processing uses hot water and steam to remove the caffeine. This process is considered safe, but it results in less flavorful coffee. The most recent method is the supercritical carbon dioxide process that removes about 97% to 99% of the caffeine.

Espresso is a concentrated form of coffee consumed plain or as an ingredient in cappuccinos, lattes, and so forth. A special espresso machine is necessary to extract the coffee. Even though its popularity is more recent, it started with the invention of a crude espresso machine in France around the early nineteenth century. Devoted espresso drinkers believe that making espresso is part of an "art" or "science" and truly believe that "extracting espresso" is not the same as "brewing coffee." Italian culture revolves around espresso and is well known for perfecting the espresso machines.

> **Americano** is espresso poured into hot water.
>
> **Cappuccino** is made by combining equal parts of steamed milk, espresso and milk froth.
>
> **Latte** is espresso mixed with plenty of steamed milk.
>
> **Mocha latte** is espresso mixed with hot steamed chocolate milk.
>
> Macchiato, Espresso Sundae, and Affogato Mocha are other modern preparations involving espresso.

Flavored coffee is a recent trend in gourmet coffee shops. Consumers add such flavors as almond, hazelnut, Irish Crème, pecan, and vanilla either directly to the beans after roasting or pour them as syrup and mix them with the already brewed coffee. Since a natural flavoring agent is hard to manage and probably expensive, many people prefer artificial flavors for convenience.

Coffee Facts

- ◆ Coffee is one of the important export commodities for twelve countries.
- ◆ Arabs were the first to brew the coffee in the mid-fifteenth century.
- ◆ In the seventeenth century, Pope Clement VIII declared it a Christian beverage.
- ◆ Dutch were the first to cultivate coffee commercially in Asia.

◆ Brazil leads the world in coffee production.
◆ The United States is a leading importer of coffee in the world.
◆ The first "instant coffee" was in a paste form.
◆ High mountain-grown coffees taste the best.

Other Uses of Coffee

The first drink made from coffee was probably a fermented alcoholic drink from the sweet pulp surrounding the seeds, a practice that still continues in Arabia today. In Indonesia and Malaya, they prepare an infusion from the dried leaves. Parts of India use the coffee pulp and parchment as manure, mulch, and cattle feed. Consumers also use roasted coffee beans for flavoring ice cream, candies, pastries, and liquor such as Kahlua. Caffeine is an additive in pain killers, stimulants, and diet pills. Coffelite, a type of plastic, is another product of coffee beans.

Chocolate

Chocolate first started as a beverage before its use in confectionary. Chocolate "beans" are the seeds of the cacao tree, *Theobroma cacao,* from Sterculiaceae. It is native to Mexico and Central America. Native Indians believed the plant was a gift of god and considered it sacred. Chocolate beans were so valuable in ancient Mexico that the Aztec Indians used them as currency; they showed their respect to the emperor by sending the beans as tribute. People in Central and South America have cultivated chocolate trees for over two thousand years. *Theobroma,* a name given by Linnaeus, means the "food of the gods" based on Aztec mythology, which believed that the cacao beans were an offer to the gods as gifts.

The Indians of Mexico prepared an esteemed beverage made from coarsely ground roasted beans mixed with corn and chili pepper. They reserved this beverage, called "*chocolatl,*" for priests and noblemen during ceremonial occasions. In 1502, Columbus was the first outsider introduced to the cacao. Later he took some of the seeds back to Europe. In 1519, the Indians offered chocolatl to a Spanish conquistador named Hernando Cortez even though the purpose of his visit was to invade Mexico. They mistakenly thought he was a reincarnation of their god Quetzalcoatl. Cortez introduced *chocolatl* to Spain in 1528 where they modified the original recipe with sugar and kept the recipe a highly guarded secret.

Spaniards monopolized the cacao business and also played an important role in spreading cacao throughout the New World. Many Central American countries began vigorously with cacao plantations. Exports to Spain soon began, and the first chocolate factory opened there. Italy and France got accustomed to the

TABLE 6.1 Non-alcoholic beverages

Beverage	Main Plant Source	Family	Parts Used	Native Origin
Asi	*Ilex vomitoria*	Aquifoliaceae	Leaves	SE United States
Chicory	*Cichorium intybus*	Asteraceae	Roots	Southern Europe
Chocolate	*Theobroma cacao*	Sterculiaceae	Seeds	South America
Coffee	*Coffea arabica*	Rubiaceae	Seeds	Ethiopia
Cola	*Cola nitida*	Sterculiaceae	Seeds	Africa
Guarana	*Paullinia cupana*	Sapindaceae	Seeds	South America
Guayusa	*Ilex guayusa*	Aquifoliaceae	Leaves	South America
Mate	*Ilex paraguariensis*	Aquifoliaceae	Leaves	South America
Tea	*Camellia sinensis*	Theaceae	Tip leaves	China; India; Tibet; Myanmar

(a)   (b)

Fig. 6.3 (a) Cacao and (b) Cacao pods
(a) © Dr. Morley Read/Shutterstock.com
(b) © tristan tan/Shutterstock.com

chocolate beverage by the early seventeenth century, and afterwards Holland, Germany, and England started using chocolate. Chocolate houses appeared and functioned as clubs alongside coffeehouses. Due to the high cost of chocolate the consumption was limited to the wealthier classes.

Processing

The fruit pod is a large, football-shaped capsule with a fleshy, thick pericarp that ripens into a yellow, red, or purple color. There are about twenty to sixty seeds ("beans") per pod, surrounded by whitish, sugary, mucilaginous pulp. A leathery skin (integument or testa) surrounds the two large, purple cotyledons (called nibs). Growers harvest the ripe pods, remove the white pulp, and allow them to ferment in heaps or in baskets. During this time the chocolate flavor develops. After fermentation, they spread the beans evenly under the sun or use artificial drying machines. Polishing the beans follows. They grade the beans by their size, weight, color, and fat content and send them to the processing centers.

Roasting the beans develops the flavor and color of the chocolate. The roasting is done at a temperature of 248° to 284°F for thirty to fifty minutes. Then the growers crack open the seeds to release the nibs. They can obtain several products from the nibs.

Products of Chocolate

The **chocolate liquor**, a dark brown oily paste, is made from crushed nibs. Merchants call it liquor only because it is liquid; it has no alcohol in it. When solidified, this paste becomes **baking chocolate**; or, when processing removes the fat (**cocoa butter**) and grinds the solids, **cocoa powder** (in 1828) forms, which has many pharmaceutical and commercial uses.

You can add cocoa butter to liquor to produce confectionary chocolate. It also appears in suntan lotions, soaps, and cosmetics.

White chocolate contains no chocolate liquor; it is mostly cocoa butter, flavorings, milk solids, and sweetener. Some believe that white chocolate is not real chocolate at all.

The recipe for **chocolate** starts with chocolate liquor, sugar, cocoa butter, vanilla, and milk. Daniel Peter of Switzerland introduced milk chocolate in 1875. **Conching** is mechanical kneading and stirring that gives a velvety smoothness to chocolate. Rodolphe Lindt invented this process in 1879.

Cacao is rich in polyphenols, similar to the ones found in green tea. The caffeine content is much lower than in coffee, tea, and cola. The alkaloid theobromine is a mild stimulant, and phenylethylamine is a slight antidepressant.

Black pod, *Phytophthora palmivora,* is one of many fungal pathogens that infect young seedlings and pods and destroys the yield. Modern research employs beneficial bio-control fungi such as *Tricoderma* to suppress the pathogens naturally.

Chocolate Facts

- ◆ Cote d'Ivoire in the western Africa leads the world in chocolate tree cultivation.
- ◆ The major producer of chocolate is the United States.
- ◆ The Swiss consume the most chocolate.
- ◆ Dark chocolates are rich sources of antioxidants.
- ◆ Growers use the criollo bean from Venezuela for the finest quality chocolates.

Cola

Cola is a name given to a variety of soft drinks made from the kola nut, ***Cola nitida*** (also referred to as ***Cola acuminata***), a member of Sterculiaceae. It is native to the forests of West Africa. The use of the kola nut, like that of coffee beans and tea leaves, has an ancient history, especially as it relates to African cultures. Local tribes chewed the seeds, called "nuts," to overcome hunger, fatigue, and to regain strength.

Kola nut trees are medium sized and thrive well in wet, humid environments. The fruit pods are up to eight inches long, have thick, leathery, mottled skin, and contain several, large red or white seeds surrounded by fleshy seed coats.

Kola nuts have a bitter flavor and contain more caffeine than coffee. A cardiac stimulant kolanin and a trace of theobromine are also present.

Outside of Africa, people have not grown kola on a large scale, except in Brazil. Growers export kola within the African continent mainly to the northern regions of Africa.

The United States was the first country to use the kola nut in a beverage—a soft drink that does not contain alcohol. In 1886, Dr. John Styth Pamberton in Atlanta, Georgia mixed powdered kola nuts and extracts from coca leaves (the cocaine plant), sugar, and caramel for color in plain water and called it **Coca-Cola**. Other ingredients, including cinnamon, vanilla, citrus flavoring and carbonated water got added to the original recipe, which is kept locked in an Atlanta bank vault. Since 1903, manufacturers have followed a modified recipe using the coca leaf extract with the cocaine removed.

Fig. 6.4 Cola
Courtesy Rani Vajravelu

Fig. 6.5 Yaupon holly
Courtesy Rani Vajravelu

Fig. 6.6 Mate plant
Courtesy Rani Vajravelu

After the successful launch of Coca-Cola other brands introduced soft drinks with slightly different formulas. Although they used kola initially, today most of these brands have switched to artificial flavorings. One soft drink that still uses kola nut is **Barr's Red Kola** made from fruit extracts. Scotland sells this beverage widely.

Cricket Cola produced in California contains kola nut extract with green tea concentrate, vanilla, and cane sugar. In 2007 in the United Kingdom, Tesco supermarket introduced **American Premium Cola** that contains kola nut extracts, spices, and vanilla.

Inca Kola is a popular Peruvian soft drink made from other plant extracts, sugar, and flavorings. Despite the name, it has no kola extracts.

Lesser Known Beverages

Mate is a social beverage made from *Ilex paraguariensis*, Aquifoliaceae, a small tree native to South America. Leaves and young twigs are steeped in hot water to make caffeinated beverage that tastes more like herbal tea than like tea or coffee. In many parts of South America drinking mate from a hollow gourd (which the name *mate* derives from) shared with friends is a common practice. Modern variations of mate include mate tea bags and carbonated soft drinks.

Another member of this family is **yaupon holly**, *Ilex vomitoria*, a species native to the southeastern United States. Native Americans brewed the small leaves of this hardy shrub to prepare a black beverage called **asi**, which is similar in color to tea from *Camellia sinensis*, and that they consumed during ceremonial rituals. The tea in large quantities or in concentrated form acted as an emetic (it induced vomiting) and a purgative. Due to the caffeine content of the leaves, a tea made from this plant was sold as a substitute for real tea during the Civil War blockade.

Guayusa, *Ilex guayusa*, is a tree related to the above two plants known for its highest percentage of caffeine content derived from the leaves. It is a tree native among the Amazon rain forests of Ecuador in South America.

Guayusa tea is prepared from boiling the leaves that can be used more than one time due to its high caffeine content. Drinking it in the early morning before sunrise is a social ritual for Ecuadorians.

Chicory from *Cichorium intybus,* Asteraceae is a hardy annual to biennial herb, native to southern Europe. When roasted and ground, the fleshy roots can substitute for coffee when real coffee is unavailable or too costly. There is no caffeine in chicory, but it produces a distinctive, roasted flavor that some people favor. Because of its pronounced taste and health benefits (believed a blood cleanser), consumers mix chicory with ground coffee powder in certain proportions. Many coffee drinkers in southeast Asia desire chicory coffee. Within the United States, Louisiana specializes in chicory flavored coffee.

Guarana is another caffeine-rich beverage prepared from the single-seeded fruits of *Paullinia cupana,* a climber from Sapindaceae. Guarana seeds are known for their highest concentration of caffeine when compared to other plant sources. The plants are native to South America and found in wild and domesticated forms. The guarana tea is made from powdered seeds. The powder prepared from roasted seeds resembles chocolate and is used to make guarana bread. When mixed with sugar, people use it for drinks and medicines. They use guarana in sweetened and carbonated soft drinks, in herbal tea, and in the capsule form. **Guarana soda** with a slight fruity taste is the national beverage of Brazil. **Guarana Antarctica**, marketed by PepsiCo, is the most famous Brazilian brand available in Brazil. The drink is also available in the United States, Portugal, Spain, and Honduras. Besides South America, consumers know guarana in the United States, France and Germany for its therapeutic properties.

Fig. 6.7 Guarana
Courtesy Rani Vajravelu

Alcoholic Beverages

Wine is fermented fruit juice from grapes, *Vitis vinifera,* a climber from Vitaceae.

One can ferment any sugary fruit to make wines, but in this section we will discuss wines made from grapes. The grape is native to the Mediterranean region and one of the oldest fruits domesticated by humans. Grapes are berries that contain natural fruit sugar, fructose, and a lot of water. Wine cultivation spread from its native origin to almost the entire world in the tropical to subtropical climates. Columbus introduced grape vines to the West Indies on his second voyage (in 1493). Within three years the Spanish started plantations in California.

Wine making started around 3000 BCE. Fine wine making still follows the traditional methods used centuries ago. Since grape wine making is a natural process, the entire process is simple. The wine makers crush the juicy fruits first. They can express juice by hand, by stomping on the grapes barefoot, or by electric

Fig. 6.8 Corn cob

© Maks Narodenko/Shutterstock. com

or mechanical presses. They run the juice mixture in closed containers to allow fermentation. To prevent microbial growth, they add sulfur dioxide. They make the white and red wines the same way except for using just the expressed juice in the former or along with the skin in the latter. If necessary, they add yeast to enhance fermentation, which continues about eight to ten days. The optimum temperatures required for the best flavored wines are 50° to 59°F for white wines and 77° to 86°F for red wines. A second fermentation follows this initial fermentation for up to a month. After allowing the sediments and dead yeast cells to settle down, the winemakers transfer the wine to aging tanks. They repeat this process across a series of tanks, a process known as **racking**. When the fermentation process is complete, they filter the remaining yeast and add sulfur dioxide to remove bacteria. During the aging process, which can range from one and a half years for white wine to up to five years for red wine, an expert called the wine master samples the wine. He or she will decide if they should bottle the wine and allow it to age further or sell it for consumption. Red wines continue to age safely for up to forty years. The quality of wine depends on factors such as the kinds of grapes, the year of harvest, and the procedures followed during fermentation. The **vintage** of a wine refers to the year in which the growers picked the grapes and initiated the wine making. **Still wines** are produced by the simple fermentation of fruit juices.

Champagne and other sparkling wines are made by adding sugar and selected yeasts to a blend of still wine before bottling. You can also identify many sparkling wines as Blanc de Blancs (from white grapes) or Blanc de Noirs (from red grapes).

Fortified wines are fermented wines to which the grower adds concentrated ethanol or a highly distilled beverage such as cognac.

TABLE 6.2 Alcoholic beverages

Family	Beverage	Main Plant Source	Parts Used
Vitaceae	Wine Champagne	*Vitis vinifera*	Fruits
Poaceae	Beer	*Hordeum vulgare*	Grain
Cannabaceae		*Humulus lupulus*	Hops
Poaceae	Whiskey	*Hordeum vulgare*	Grain
	Vodka		Grain
	Gin		Grain
	Rum	*Saccharum officinarum*	Stem sap
	Sake	*Oryza sativa*	Grain
	Chicha	*Zea mays*	Grain
Agavaceae	Pulque	*Agave atrovirens*	Stem sap
	Tequilla	*Agave tequilana*	Stem
	Mezcal	*Agave palmeri*	Stem

Hop plant
© photofriend/Shutterstock.com

Agave plant
© holbox/Shutterstock.com

Beer is made by brewing a mix of malt, hops, and water. Malt is the sprouted grain of barley, *Hordeum vulgare* (Poaceae); hops are the female inflorescence of *Humulus lupulus,* Cannabaceae. Most U.S. breweries rarely add **adjuncts**, which are the unmalted grains, such as barley, wheat, and rice, along with corn grits, corn syrup, and other carbohydrates such as potatoes.

Whiskey is made using similar procedures except the producers use and distill malted barley alone or with other grains. Manufacturers make **vodka** and **gin** with malted barley except they have a very high percentage of alcohol. **Rum** is the fermented molasses or juice of sugarcane, *Saccharum officinarum,* Poaceae.

Sake is made from fermented rice grain, and usually served heated. Rice, *Oryza sativa,* Poaceae, is the main ingredient along with such fermenting agents as *Aspergillus* and yeast. The refined rice is cooked, cooled and allowed to ferment in fungus and water mixture, which is later squeezed out and allowed to mature for about forty days. Even though the preparation is more like a beer, people consider it a rice wine and a traditional beverage of Japan.

Chicha is a traditional beverage in Central and South America. The chewed corn kernels are mixed with ground corn, *Zea mays* (another member of Poaceae), and water. Saliva mixed in the chewed corn provides the enzyme amylase needed to break down the starch. Modern methods skip the chewing practice.

Pulque, tequila, and **mezcal** are prepared from the fleshy stems of *Agave atrovirens, A. tequilana* and *A. palmeri* respectively of Agavaceae. To make pulque, the producer collects the sweet sap by puncturing the stems and allowing it to ferment for about fourteen days. The other two are beverages derived from the cooked, fleshy stems ("hearts"), which are allowed to ferment and later distilled. People in Mexico widely consume these alcoholic beverages.

Review Questions

Give brief answers in the given space.

1. List all caffeine beverages and their plant sources.

2. Describe the events that led to the Boston Tea Party.

3. Besides Camellia, what other plant genera give us leaf-based tea?

4. List all alcoholic beverages and their plant sources.

5. Describe how coffee spread to the rest of the world from Africa.

6. List all the beverages derived from the members of Aquifoliaceae.

7. Which countries lead in the cultivation of coffee and chocolate?

8. What is espresso? List other products that use espresso.

9. Describe the preparation of rice wine in Japan.

10. Distinguish between the two processing methods of coffee beans.

Multiple Choice Questions

For each question, choose one best answer.

1. Who invented the method of brewing coffee for the first time?
   A. Arabs
   B. Americans
   C. Europeans
   D. Chinese
   E. Ethiopians

2. One of the steps in tea processing in which the fermented leaves undergo a second drying process is _____.
   A. firing
   B. blowing
   C. withering
   D. oxidation
   E. rolling

3. Pope Clement VIII declared this beverage a "Christian Beverage." Which one?
   A. Chocolate
   B. Coffee
   C. Green Tea
   D. Oolong tea
   E. None of the above

4. Traditional preparations of rum used:
   A. barley malt
   B. cooked rice
   C. sugarcane molasses
   D. potato
   E. hops

Match the following chemical compounds to the appropriate sources. You may use a choice once, more than once, or not at all.

Column 1	**Column 2**
5. Theophylline	A. Cola leaves
	B. Cacao seeds
6. Kolanin	C. Tea leaves
	D. Cola seeds

7. Which of the following did people use as an alternative beverage during the Civil War blockade in the United States?
    A. Chicory
    B. Ginger
    C. Green tea
    D. Yaupon holly
    E. Both A and C

8. Caffeine is _____.
    A. a stimulant
    B. an alkaloid
    C. found in seeds or leaves
    D. all of the above
    E. only A and B

9. Name the geographic area that currently leads in cacao cultivation.
    A. South America
    B. China
    C. Africa
    D. Islands in Indian Ocean
    E. Australia

10. Sterculiaceae is the family of which of the following?
    A. *Cola*
    B. *Theobroma*
    C. *Thea*
    D. All of the above
    E. Only A and B

True/false Questions

1. One obtains "hard coffee" by pulping the coffee berries using the dry method.   **True**   **False**

2. A spicy beverage from cacao was first prepared by Ethiopians.   **True**   **False**

3. Different kinds of tea, including green tea, developed in China during the Ming Dynasty.   **True**   **False**

4. Yaupon holly plants are native to Mexico.   **True**   **False**

5. Brazil is the leading producer of kola seeds.   **True**   **False**

6. Chocolate liquor consists of ground seeds mixed with alcohol.   **True**   **False**

7. Besides caffeine, the stimulating effect of tea is due to tannin.   **True**   **False**

8. Chicory is a coffee substitute because of its caffeine content.   **True**   **False**

9. Guayusa leaves have the highest caffeine content for any plant known so far.   **True**   **False**

10. In beverage industry, pea berry refers to a single-seeded kola nut.   **True**   **False**

CHAPTER 7
VEGETABLE OILS

One of the organic compounds found in plants is fats. Oils and waxes are two major types of plant fats. Fruits and seeds store oil to provide energy for germinating embryos. Certain plants store more oil in their seeds than other plants that store food mainly as starch or protein. Waxes are fats that appear as a waterproof coating on plant leaves and fruits. These two plant fats provide a number of commercial products, including cooking oils, car wax, varnish, and more, as discussed in this chapter. Most vegetable oils are edible like peanut oil and soybean oil, but there are some inedible oils such as tung oil, which we use in paints, polishes, and medicines. The focus of this chapter is to discuss plant sources and extraction methods, as well as the indigenous and modern uses of vegetable oils.

Vegetable oils play an important role in human nutrition. Since ancient times, humans have learned to extract such oily storage from plants by employing a variety of methods. When added to food, oils distribute heat evenly, enhance their flavor and texture, and provide more calories than any other storage food from plants. People also favor vegetable oils for a healthy eating lifestyle because they have no cholesterol and no trans fats by themselves unless misused and modified by humans.

Oils are liquid at room temperature and composed of triglycerides that are long-chain organic compounds. A triglyceride is a combination of one glycerol and three fatty acid compounds. The type of fatty acids consumed in a daily diet can influence the overall health of a person. In order to distinguish the healthy oils from others, let us see how they are classified.

Types of Fatty Acids

There are about forty kinds of fatty acids that occur naturally. Researchers classified them based on the presence and number of double bonds that exist in the fat's molecular structure.

Saturated fatty acids are vegetable oils without double bonds. These include palmitic (palm oil) and capric acids (coconut oil).

Unsaturated fatty acids are vegetable oils with double bonds. These include oleic acid (olive oil) and linoleic acids (soybean oil).

> Monounsaturated fatty acids have one double bond and are the least injurious to human health. *Example:* Oleic acid in olive oil.

> Polyunsaturated fatty acids have more than one double bond. *Example:* Linoleic acid in soybean oil.

Essential fatty acids (EFAs) are necessary fats involved in biological processes that humans cannot synthesize. Omega fatty acids are unsaturated fatty acids with high carbon lengths. They reduce triglyceride levels in the blood, reduce blood pressure, and may have anti-carcinogenic properties. EFAs support the cardiovascular, nervous, immune, and reproductive systems. You must obtain EFAs from unsaturated fatty acids in the diet, such as linolenic, linoleic, and oleic acids. Some vegetable sources are flaxseed, hemp, and canola oils, as well as the seeds of pumpkins, sunflowers, and walnuts along with leafy vegetables.

There are two families of EFAs: Omega-6 derived from linoleic acid (soybean and safflower oil) and Omega-3 derived from linolenic acid (sesame and flax seed oil). An ideal intake ratio of Omega-6 to Omega-3 fatty acids is between 1:1 and 4:1.

Hydrogenation is a process of converting unsaturated fats to solids with the addition of hydrogen. The amount of added hydrogen can control the degree of softness and the melting point. This discovery has resulted since 1900 in vegetable fats and shortenings that substitute for animal butter. **Trans fatty acids** are partially hydrogenated fatty acids. Foods high in trans fatty acids (margarine, shortening, and commercially fried food) tend to raise cholesterol levels in the blood. To ease the health concerns related to these hydrogenated fats, modern technology has succeeded in making fats that intestinal walls cannot absorb like they do other fats. Olestra™ by Proctor & Gamble and Promise™ are fat substitutes that produce the desired taste in food items.

Oil Extraction Methods

Plant seeds and fruits are loaded with oils that may not be readily visible. Such oil-containing tissues must be crushed to release the trapped oils. Traditionally, people used heavy **grinding stones** to crush the oil-containing tissues to release the oils. They suspended the crushed mass of tissue and oil in a cloth filter to separate the dripping oil.

Modern methods vary based on the oil source of the plant. Oily mesocarp, such as in olives, get processed in ways that are different than the processing methods employed for oily cotyledons, endosperms, or embryos. We describe the standard methods below.

One of the old oil extraction methods was to use a large mortar and pestle device fixed deep into the ground for added stability. The pestle attaches to a manually driven revolving shaft that uses bulls. After a few hours of grinding, oil separates out, and scooped out from the remaining cellular mass. Such a cold press method is still prevalent in remote parts of India and on adjacent islands in the Indian Ocean where local oil merchants obtain up to four to five gallons of oil at one time. This takes about three hours of intensive labor from two people, one to manage the crusher and the other to drive the bulls in constant circles to control the shaft holding the pestle. In Tamil Nadu, one of the southern states of India, the people call these devices **chekku**. Such wooden chekku used to be a common device a decade ago, but they are fast disappearing and giving way to modern machine presses called expellers. Before the invention of **expellers**, people crushed large amounts of oil seeds using cold or hot pressing. Cold pressing yielded pure oils compared to hot pressing where they applied heat to extract more oil.

The invention of the modern screw press (expeller) that generates high heat occurred in 1900. The expeller allows for continuous feeding of the seeds at one end of the device. The turning screw allows the crushed debris and pressed oil to move out into their separate slots and collect elsewhere. The crushed seed residue still may contain a small percentage of oil. Another way to extract as much oil as possible is to use a chemical solvent method using hexane and heat.

Due to health reasons involved with the use of chemicals, many companies opt for initial expeller pressing even though it yields only about 70% of the oil from the seeds. Because of this, expeller oil is more expensive.

Prior to oil extraction all oil sources are cleaned and dried. Seed coats surround the seeds, as in the peanut, and a fruit wall surrounds sunflower seeds. **Husk** is the name used to refer to these seed coats and/or fruit walls. The next step is **flaking**, followed by **conditioning** that allows for cooking the broken pieces using high heat.

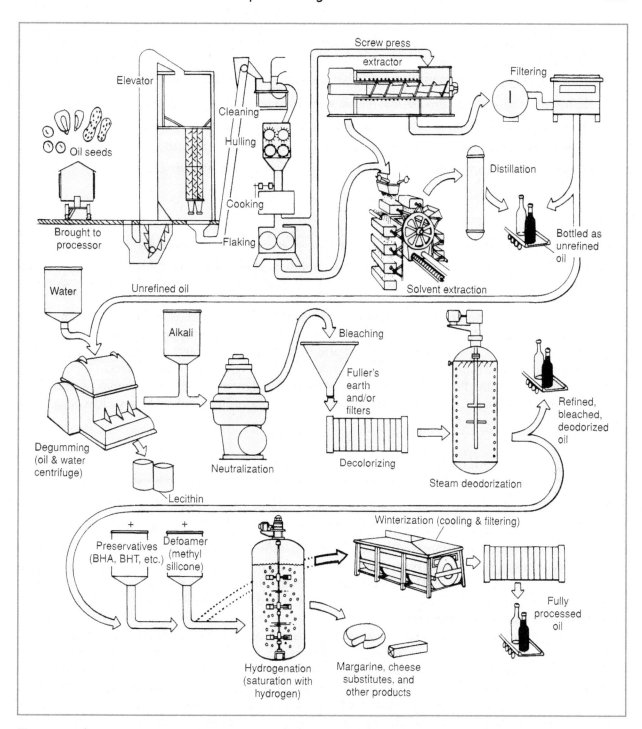

Fig. 7.1 A diagrammatic representation of the extraction and processing of vegetable oils. Top left: Hulling and flaking of the whole seeds. Top right: Expeller pressing and solvent pressing of the prepressed seeds. Center: Refining, degumming, and bleaching of the extracted oil. Bottom: Hydrogentation and winterization processes.

From Economic Botany: Plants in our World 3E by Molly Conner Ogorzaly, 2000. Reprinted by permission of The McGraw-Hill Companies.

Such prepared seeds are initially pressed and separated from the residue or sent through solvent extraction. The residue called **meal** is heated with steam to remove any remaining toxic chemicals (in the case of solvent extraction) and sold for cattle feed.

Most oils undergo refining to make them usable in various cooking methods.

Refining is a neutralizing process that removes any free fatty acids; it uses alkali.

Degumming is removing any mucilaginous material in the oil by centrifugation.

Bleaching is the removal of pigments from the oil using diatomaceous earth.

Deodorizing is removing any strong odor under a vacuum, and **winterizing** is removing the clouding ability of the oil by chilling.

Smoking Points and the Health Connection

In cooking, it is important to select the oil based on its smoking point. Oils with a high smoking point will withstand high cooking temperatures and are preferable in grilling and deep frying. Refined oils extracted using heat normally have high smoking points. Some examples are peanut oil and rice bran oil.

Oils with low smoking points will begin to smoke when the temperature exceeds the optimum level. That can result in a poor acrid flavor and bad tasting food, and it may pose a risk for fire. The smoke may contain carcinogenic compounds. Unrefined oils extracted using a cold method like olive oil and other nut oils have a lower smoking point and come recommended for sautéing, stir frying, and in salad dressings.

One can refine any unrefined oil and create a less flavorful, sediment free oil. Some brands of refined oils can have lower smoke points. Some of the best refined oils suitable for high heat cooking such as deep frying are peanut, canola, safflower, and soy.

Some Well-Known Vegetable Oils

Canola Oil

Canola oil is from the seeds of the plant called rape, *Brassica napus,* or turnip, *B. campestris,* both from Brassicaceae. Its original name, rapeseed oil, has changed to reflect the Canadian origin of the oil. Rapeseed oil had limited cooking use in olden times due to its sulfur compounds and a fatty acid called erucic acid that imparted an unpleasant acrid taste. The modern cultivars from Canada (where the name canola derives from) have reduced unpleasant odors in part due to preheating techniques used to remove the enzymes responsible for the breakdown of sulfur compounds.

The seeds of canola are very similar to mustard seeds, that is, small and borne in fruit pods called siliques that mature starting from the basal portion of the inflorescence. To avoid losing the seeds by shattering fruit pods, the farmers use machines and harvest them before the entire seed pods mature. They then allow the seeds to mature in the pods. The seeds then are threshed, cleaned, and taken for crushing to extract the oil.

Canola oil has high levels of monounsaturated fatty acids, a low level of saturated fatty acids, and moderate amounts of polyunsaturated fatty acids. It has precursor compounds for omega-3 and omega-6 fatty acids and a high smoke point that makes it suitable for deep fry cooking.

In addition to cooking uses, canola oil has industrial uses as a lubricant as well as in other situations involving high temperatures. Another cultivar with high levels of erucic acid is preferred for this purpose.

Corn Oil

Corn oil is from a monocot plant known as maize, also called corn, *Zea mays,* Poaceae. They are native to Mexico and Central America. Corn plants are tall, annual plants with unisexual inflorescences. A corn ear is the female inflorescence covered by the elongated styles known as silks. Once mature, each ear will hold about three hundred or more kernels that are usually yellow but that may vary from brown to red to bluish-gray and black. The corn kernels are the grains or caryopsis, each with a well-developed starchy endosperm and a small embryo. Corn oil is a byproduct of the corn-milling industry that began in the United States during late nineteenth century.

Processers extract oil from the scutellum portion of the **germ** (embryo) in the grain and use the rest of the grain area, mainly the endosperm part, for starch. Farmers raise several varieties of corn for flour, oil, protein, sugar, and for popcorn. Before extracting oil from the germs, they must mill the grain to remove the germ from the rest of the grain. They do it either by wet or dry method. In both methods, they remove the lighter embryos by slight friction, later conveying the germs to a dryer and to an oil recovery facility. Solvents separate a crude corn oil that later becomes refined.

Due to its high smoke point, consumers value refined corn oil as frying oil. It has a mild taste, a high amount of antioxidants and polyunsaturated fatty acid, and a moderate amount of mono unsaturated fatty acid. Most of the oil gets used as salad oil and in shortenings. Other industrial uses of corn oil include using it in corn plastics, paints, textiles, insecticides, and pharmaceutical products. In recent years people have valued corn oil as a source of biodiesel.

Cottonseed Oil

Cottonseed oil is from cotton plants, several species of *Gossypium,* Malvaceae. It is one of the four genetically modified plants, including corn, soy, and rape (canola). Commercial exploitation of the seeds for oil is recent; however the practice was common in India since ancient times. Oil extraction was by pounding and boiling the seeds. They used the oil in medicine and as lamp oil. A compound called **gossypol**, a naturally occurring toxin, imparted a bitter taste to the crude oil, so they did not use the oil for cooking, but rather as an insecticide. They used the seeds as cattle feed and fertilizer. But after the early 1900s the perception of cotton seed oil changed. It became the original American vegetable oil. Its bland flavor and mild nutty taste became a scale to compare with other oils and became preferred in many fried items for retaining crispness. This includes potato chips and other snack items.

The change in its reputation was because of **David Wesson** who tried a process combining caustic soda, deodorant, and steam to eliminate the bitterness. He succeeded in making edible cooking oil marketed as Wesson Oil in 1900. Proctor & Gamble invented the first American vegetable shortening called **Crisco** in 1911 by the hydrogenation method of cotton seed oil. Cottonseed oil contains no trans fatty acids and no cholesterol. It has both omega-6 and omega-3 fatty acids, but the former is in a high percentage, which makes the ratio unhealthy.

Before the cottonseeds are crushed for oil, machines remove the fine trichomes of cotton seeds called **linters**. Initial crushing yields dark-colored, semi toxic oil. Refining by filtering, bleaching, and deodorizing results in a tasteless, colorless oil. Such refined oil is useful in the liquid form or hydrogenated to make margarines and other shortenings.

The removed linters are useful in making cellulose fibers like rayon and also in papermaking. The cottonseed cake that remains after pressing the oil out gets used as cattle feed.

Olive Oil

Olive oil is from *Olea europaea,* Oleaceae. Chapter 4 discusses this plant, its fruit and history. The oily mesocarp is the source of olive oil. The ancient Greeks considered olive oil the "nectar of the gods." Theophrastus called it "the great therapeutic." Egyptians used olive oil as a cleanser, a moisturizer, as medicine, as lamp oil, and in religious practices. Researchers also believe that workers used olives and olive oil in the construction of the Egyptian pyramids. The Mediterranean region leads in the production of olives.

Fig. 7.2 Peanut products
© govindji/Shutterstock.com

To extract the oil, processers crush the pitted olive pulp to make a paste. They then slowly mix and churn (a process called **malaxation**) this olive paste until the oil droplets release and accumulate. They may apply heat to increase the oil yield, taking precaution to avoid oxidation of the oil that would interfere with the overall flavor and shelf life of the oil. Traditionally, they would separate this churned paste from the oil with a large heavy press. Modern methods include a phase separator that uses a decanter centrifuge to separate the oil. **Pomace** is the remaining solid substance with very little oil after the extraction of the oil droplets. This pomace yields low grade oil after it goes through a combination method of heat and solvent extraction.

The first press yields the superior quality oil. Subsequent presses yield oils of inferior quality. In retail business, merchants classify different grades of olive oil based on the standards set by The International Olive Oil Council (IOOC) based in Madrid, Spain. The IOOC is an intergovernmental organization that promotes the quality and production of the olive among its twenty-three member nations. The United States is not a member of the IOOC.

Extra-virgin olive oil is the cold pressed oil with superior taste and a lower acidity less than 1%. **Virgin olive oil** is the next grade with a little more acidity (<2%) and a good taste.

These two oils are pure extracts and may not contain any refined oil.

A blend of the above two and of refined oil is **pure olive oil**. A blend of virgin oil and refined olive oil has the label **olive oil**, which tastes milder than all of the above oils.

Olive oil is rich in monounsaturated fats and considered one of the healthy oils as long as consumers use it in moderation. They prefer the oil in salads, sautéed, and stir-fried, not for frying due to its low smoke point. It is undesirable in baked desserts due to its strong flavor. Manufacturers use olive oil in a range of cosmetic and skin care products, and the soap and pharmaceutical industry have other uses for the oil. Italian and Spanish olive oils are well-known within the United States. Spain provides a large part of the olive oil consumed in the United States. Extra virgin olive oil from Spain and Italy are in great demand.

Peanut Oil

Peanut oil is obtained by crushing the oily cotyledons of a legume plant, *Arachis* **hypogaea**, Fabaceae. Use of the seeds for oil is relatively recent compared to other oil sources. For a detailed description of this plant, please see Chapter 5. A French firm first crushed peanuts to make peanut oil during the late 1800s. Manufacturers shell and dry the peanut seeds before they take them to mills for crushing and for solvent extraction. The United Kingdom and most of Asia know peanut oil as **ground nut oil**. It is rich in monounsaturated fats with a medium high smoke point, making it a good choice for light sautéing, sauces, and deep frying. It does not retain flavors. Peanut oil is rich in palmitic acid, oleic acid, and linoleic acids. Fully refined peanut oil with all the proteins removed is considered safe for people with allergies. However, one should be wary of oil extraction methods used in such oils. It used to be the original source of fuel for the diesel engine. Consumers use the peanut cake as a feed for livestock.

In the United States, people use only about 25% of the crop for oil; they use the rest to make snacks, most of which are peanut butter.

Rice Bran Oil

Rice bran oil is a recent addition to the health-conscious modern public. The source of oil is *Oryza sativa*, Poaceae. Japanese women have used this oil as a moisturizer for a long time throughout history. Because of their shiny, smooth skin—apparently the result of their regular use of rice bran oil—these women received the nickname of "**bran beauty**." Asian cultures, including Japan and China, have used this oil for a long time in snack foods and value it as ideal for deep frying and other cooking needs.

Manufacturers extract rice bran oil from the bran portion of the seed, which includes the **germ** (embryo) of the rice grain.

In the United States, manufacturers have produced it commercially since 1994, and upscale restaurants and specialty snack preparations are currently using it. Consumers value it for the nutritional benefits it offers as it has a rich amount of mono- and polyunsaturated fats, vitamin E group antioxidants, oryzanol, tocopherol, and tocotrienol. Researchers report that oryzanol reduces LDL levels in the blood without reducing HDL levels. Tocotrienol is a powerful vitamin E, known for its anti-cancer effects. Rice bran oil has the highest amount of total antioxidants, followed by soybean oil.

Rice bran oil gives a desirable color, texture, and delicate flavor to food without leaving an aftertaste. Consumers most value this oil's nutty flavor and its stability from turning rancid. Due to high smoke points, they favor it in high heat grilling. It is ideal for making shortenings and margarines, rendering a smooth spread and taste. In addition the cosmetic industry values it, especially in sunscreen products.

Safflower plant
© ncristian/Shutterstock.com

Safflower Oil

Safflower oil is from the seeds of a spiny annual called safflower, *Carthamus tinctorius*, Asteraceae. Its native area is uncertain, but the eastern Mediterranean region domesticates it widely. The ancient Egyptians used to obtain red and yellow dyes from this plant, and the Indians later extracted the oil, which they mainly used in medicine. People only discovered its value as an oil during the last thirty or forty years. Two varieties of plants produce two types of oil—one highly monounsaturated, the other highly polyunsaturated—along with two types of unsaturated fatty acids. Consumers use the first type of oil, which is heat stable, in various ways, ranging from infant food formulas to frying snack items; they use the other type in salad oils and soft margarines.

The oil is similar to sunflower oil, nearly colorless, extremely mild, and used in salad dressing and cooking. A few decades ago, safflower oil was prized for its use in paints and varnishes in the United States. Due to its high fiber and moderate protein content, safflower meal is valuable in livestock and cattle feed. Whole seeds are used as bird seeds.

Sesame Oil

Sesame oil is seed oil from *Sesamum indicum*, Pedaliaceae. People in the United States do not much prefer it, other than in Korean and Southeast Asian menu items. It appears more ancient than linseed oil, but concrete evidence is lacking. India, Africa, Middle East, and China consume the most sesame oil. Ethiopia appears to

Sesame plant

Fruit l.s.

Fig. 7.3 Sesame plant and fruit
© homi/Shutterstock.com
© Swapan Photography/Shutterstock.com

be the region of native origin for this oil crop. Sesame oil was rare in the Mediterranean area before 600 BCE and used only by the wealthy. Known as **gingelly oil** or **til oil** in India, historians have documented its use since early times. Hindus consider this oil sacred and use it to light lamps in their prayer areas. Some south Indian names for this oil translate to "good oil" (nalla ennai), attesting to its value in food and health. In India people use sesame oil in body and hair massages, as well as a constituent in Ayurvedic drugs and cosmetics. Sesame oil has a reputation as healthy oil both in topical and internal administration. It has a myriad of local uses in promoting strength and vitality, reducing aches and pains, as an excellent moisturizer and a laxative, and, in general, for slowing the aging process. Unfortunately researchers have not proven any of these claims scientifically yet. The introduction of other refined oils has reduced its use in cooking and frying. Tahini is a sesame-based paste made from hulled, lightly toasted seeds popular in Middle Eastern cooking. Chinese, Korean, and Japanese dishes use a similar paste made from unhulled sesame seeds.

Sesame plants are herbaceous annuals with small, funnel-shaped flowers that are white or pinkish in color. The tender plant parts and flowers contain mucilage. The fruit pods are flat capsules that one must dry and tap to remove numerous flat, oval seeds. There are two wild varieties, one with white seeds and the other with black seeds.

Consumers can press the dried seeds raw (cold press) or toast them prior to that. The cold press yields light brown, mild flavored oil compared to toasted seed oil that yields dark brown, strong flavored oil.

Dried seed oils are mixed with a small amount of molasses or dry palm sugar to help release the oil from the seeds. Alternately, time-efficient modern rotary machines press the oil in huge amounts. Manufacturers simply filter pressed oil to remove debris and directly store it in containers. Sesame oil has a high smoking point and retains its natural structure and flavor even after heating it to a high temperature. Because of this stability, sesame seeds are preferable in pickled items stored at room temperature for long periods of time. Sesame oil is known for its vitamin E content, an antioxidant correlated with lowering LDL levels in blood. The oil is a rich source of polyunsaturated fatty acids, omega-6 fatty acids, and two naturally occurring preservatives, **sesamol** and **sesamin**.

Soybean Oil

Soybean oil is from a bean plant, soybean, *Glycine max*, Fabaceae. This is a small, prostrate, annual plant, native to East Asia. Soybean is an important global crop; besides oil, it provides a high quality protein. The fruit pods resemble most bean pods called legumes. The source of oil is the seed that occurs in various sizes and colors ranging from brown, yellow, green, black, and mottled. Before pressing for oil, manufacturers remove the seeds from the pods, a process known as **shelling**. They clean and heat-treat the seeds to remove fungal pathogens and to denature certain enzymes that interfere with protein digestion. The seeds contain 40% protein and 20% oil.

Europeans have been pressing the oil since the 1700s. In the United States, farmers initially grew soybeans for hay. Starting around the 1950s the United States increased its cultivation of soybeans mainly for the oil, and this reached its peak by the 1980s. The soybean is a main source of oil in the world compared to all other vegetable oils. The United States is the leading producer of soybean oil.

Manufacturers crack the oil-filled seeds before the oil is extracted using solvents. They use the remaining cake that is rich in protein as animal feed. Lecithin is an additive obtained as a by-product of the oil industry that companies use as a stabilizing agent in cosmetics, food, leather, and plastic. The pharmaceutical industries also use it. Soybean oil is commonly called "vegetable oil." It is low in saturated fat and high in polyunsaturated fatty acids and linoleic acids. It also has a moderate amount of monounsaturated fat. It has a high smoke point, which makes it suitable as frying oil. Besides cooking, consumers use the oil in salad dressings, shortenings, non-dairy milk, creamers, and spreads. Soybean oil has industrial uses in the manufacture of paints, inks, soap, insecticides, disinfectants, linoleum, and many other products.

Sunflower Oil

Sunflower oil is seed oil from *Helianthus annuus*, Asteraceae. Native to North America, wild populations are abundant in Mexico, the United States, and Canada. Fossil remains of sunflower fruits from 800 BCE were found in eastern North America. There are evidences to support the use of the seeds as food and oil among Native Americans.

Sunflower plants are tall, annual, or perennial plants with large, ovate leaves. The **florets** (tiny flowers) are aggregated in a large, flat inflorescence called a **head** or a **capitulum**. The head has two types of florets: brown, disc florets in the center surrounded by large, yellow, strap-shaped ray florets. The fruit is an achene, and the single seed inside is the oil source. There are two types of sunflower achenes, based on size and color: larger, gray, and striped ones used as snacks, and smaller, black ones used for oil. Selected cultivars with larger heads and oil-yielding dwarf varieties are recent additions to the sunflower industry. Crude sunflower oil is used without processing or mixed with diesel to operate farm machinery in the United States. Due to clouding properties, manufacturers winterize and moderately refine sunflower oil used in cooking.

Fig. 7.4 Sunflower plant
© maxpro/Shutterstock.com

Sunflower oil is light in taste and appearance, which consumers prefer in cooking, frying, and as salad oil. Several types of oils with different percentages of linoleic, oleic, and stearic fatty acids are produced based on the climate and the variety. This oil is a combination of mono- and polyunsaturated fatty acids with a high amount of vitamin E. People also use it as an emollient, and it has applications in the cosmetic and paint industries.

Coconut Oil

Coconut oil is extracted from the plants of *Cocos nucifera,* Arecaceae. Chapter 4 described the fruits. Coconut oil is extracted from the oily endosperm of the single seed. Archaeological evidences suggest that Polynesians probably extracted oil in ancient times.

In order to extract the oil, the entire mature coconut is cleaned out to remove the fibrous outer layers leaving the endocarp shell. This shell (or **pit**) is broken in half and dried until the cellular endosperm (**copra**) separates from the endocarp shell. Sometimes they are separated out manually. For the drying process, either direct sun or kilns are used. At this point, the copra is hard and rich in oil. It is broken in to smaller pieces and simply crushed in a heavy machine to extract the oil. This crude coconut oil is colorless and has a mild, sweet fragrance of coconut. This oil is then bleached and deodorized to obtain refined oil that is milder with a higher smoking point.

Another way of oil extraction is straight from the coconut milk obtained from fresh cellular endosperm meat. In a process called **wet-milling**, oil is separated from the water by boiling, fermentation, and mechanical filtration. Such oil is called **virgin coconut oil**, even though its standard is not comparable to olive oil and no organization monitors it.

Recent trends in nutrition lead towards the use of virgin coconut oil due to a high percentage of beneficial saturated fats and medium chain triglycerides. Unrefined coconut oil is stable and slow to oxidize and turn rancid. It solidifies below 75°F in to solid white fat called coconut butter. Refined coconut oils are more suitable for frying than unrefined oils. However, it is strongly recommended that a small amount of coconut oil is used as a substitute for another saturated fat in your diet.

Castor Oil

Castor oil is obtained from the seeds of *Ricinus communis* of Euphorbiaceae. Chapter 10 describes this plant. The species is native to warm regions of Africa. The oldest records are from Egypt dated 4000 BCE. India is the world's largest producer, followed by Brazil.

Fig. 7.5 Castor bean plant
© Eugene Sergeev/Shutterstock.com

The fruits of castor plants are spiny capsules, three-lobed, and each with a single, mottled, attractive seed called a bean. Each has a well developed oily endosperm, the source of oil.

Except for the refined oil, other parts of the fruit are extremely poisonous. Three poisonous compounds found in the seeds are the alkaloid called ricinine, a protein called **ricin**, and a protein-polysaccharide named CB-A. Entire seeds and the hull are loaded with these compounds. Researchers have developed recent varieties that lack poisonous compounds. Consumers do not use castor oil as cooking oil; rather, they use it variously as noted below.

People used cold pressed castor oil for skin conditions, including burns, cuts, abrasions, lesions, and eruptions. They also used it as a rub to ease abdominal pain and as a laxative, as well as to treat headache, muscle pain, and inflammation. For medicinal uses, they dry and press the seeds to extract oils at a lower temperature (<122°F) to keep the ricin. Ancient uses were as lamp oil, a purgative, in medicine, and in religious ceremonies.

Castor oil is pale yellow, sticky, and viscous with a mild unique taste. It is rich in monounsaturated fats with a high percentage of ricinoleic acid, followed by oleic and linoleic acids. People use a food grade castor oil as flavoring, as an additive, and in candies. Castor oil is used as lubricants, paints, inks, dyes, waxes, polishes, synthetic polymers, and in cosmetics. We know hydrogenated castor oil as **castor wax** that is insoluble, brittle, and hard. It is used in coatings and greases that require resistance to moisture and other chemicals.

Neem Oil

Neem oil is pressed from the seeds of *Azadirachta indica,* Meliaceae. This is a plant native to India and widely grown in adjacent tropics. The medium-sized trees have compound leaves. The fruits are oblong, yellow drupes about one inch long. The flesh is cream-colored and soft, eaten by birds that leave out the pits. The dried pits are shelled to remove the oily seeds and crushed through cold pressing to yield the highest quality oil for use in medicine. The solvent-extraction methods yield lower quality oil used in soap manufacturing. Farmers use neem cake, obtained from the remaining seed tissue, as fertilizer and insecticide. Neem oil had use as lamp oil in ancient India.

Due to its strong smell and bitterness rendered by triterpenoid compounds, consumers do not use neem oil in cooking. It has applications in Ayurvedic medicinal practices to treat a multitude

Fig. 7.6   Neem leaves
© ARTEKI/Shutterstock.com

of health conditions, including leprosy, malaria, and tuberculosis; and in common folk medicine to treat skin conditions, fever, and rheumatism, as well as in contraception. Because of its wide range of applications in medicine, people in ancient India considered the neem tree the healer of all ailments. The cosmetic industry uses neem oil in skin sprays, lotions, soaps, toothpaste, and cream. Neem leaves also have antiseptic and insecticidal properties. Because of this, every traditional household in south India grows at least one neem tree outside of its house. Organic farmers use neem oil-based formulations widely as a natural pesticide.

Additional Sources of Oils and Waxes

Nuts are generally rich in oil content, and several so-called nuts, such as walnut and almond, provide oils used in cooking or as flavoring. The mesocarp of avocado is rich in oil content that one can extract to make avocado oil. We obtain palm oil and palm kernel oil from a single source, *Elaeis guineensis,* where the oily mesocarp and the oily seeds are separately squeezed to extract two types of oil. In certain other plants, the waxy coating surrounding the fruits or leaves are scraped off the surface and used in commercial products, such as candles and car wax. A short list given below includes a few such sources not discussed in this chapter so far.

TABLE 7.1 Some additional sources of oils and waxes

Name and Type	The Source of Oil	Binomial	Family	Applications
Almond oil—monounsaturated	seeds	*Prunus amygdalus*	Rosaceae	In healthy cooking practices
Avocado oil—monounsaturated	mesocarp of drupe	*Persea americana*	Lauraceae	Used in healthy cooking
Grape seed oil—polyunsaturated	seeds	*Vitis vinifera*	Vitaceae	Salad oil; massage oil; in cosmetics, cooking
Linseed oil—polyunsaturated	seeds	*Linum usitatissimum*	Linaceae	Applied as a glaze on wood and in paints
Palm oil and Palm kernel oil—saturated	mesocarp of drupes and seeds	*Elaeis guineensis*	Arecaceae	Limited use in healthy cooking; soap, candles, shortening
Tung oil—polyunsaturated	seeds	*Aleurites fordii*	Euphorbiaceae	In paints, conditioner for wood, in India ink
Walnut oil—polyunsaturated	seeds	*Juglans regia*	Juglandaceae	Baking, sautéing; rich in omega-3 fatty acids
Carnauba wax	young leaves	*Copernicia cerifera*	Arecaceae	Car wax, shoe polish
Bayberry wax	berries	*Myrica pensylvanica*	Myricaceae	Fragrant candles

Review Questions

Give brief answers in the given space.

1. Differentiate saturated and unsaturated fatty acids found in vegetable oils.

2. What are essential fatty acids? Cite at least two examples.

3. Describe the ancient oil extraction methods used prior to the invention of expellers.

4. What is the smoking point of oil? Discuss its role in food quality and health.

5. List three plant oils that are rich in monounsaturated fatty acids.

6. What is refined oil? How does it influence the use of certain oils in cooking?

7. Explain the differences between various grades of olive oil.

8. Why is castor oil not suitable for human consumption?

9. Describe the methods involved in cottonseed oil processing.

10. What is rice bran oil? What are its applications to human society?

Multiple Choice Questions

For each question, choose one best answer.

1. Which of the following vegetable oils is derived mainly from the germ of the grain?
    A. Canola oil
    B. Corn oil
    C. Rice bran oil
    D. All of the above
    E. Only B and C

2. Which of the following has a poisonous seed that yields non-toxic oil?
    A. Sesame
    B. Rape
    C. Castor
    D. Corn
    E. Palm

3. Which cold pressed oil do we know for its stable properties that also is a preservative in Indian pickled items?
    A. Neem
    B. Peanut
    C. Coconut
    D. Sesame
    E. Soybean

4. Which is the plant source for canola oil?
    A. Corn
    B. Cottonseed
    C. Safflower
    D. Sunflower
    E. Rapeseed

5. Which oil-yielding plant also had ancient uses as a source of yellow and red dyes?
   A. Sunflower
   B. Olive
   C. Castor
   D. Safflower
   E. Sesame

6. Which of the following pairs of oils do we press out of the endosperm?
   A. Soybean and peanut
   B. Olive and palm
   C. Sunflower and safflower
   D. Ricebran and corn
   E. Coconut and castor

7. Which of the following vegetable oils has the highest percentage of saturated fatty acids?
   A. Ricebran oil
   B. Coconut oil
   C. Sesame oil
   D. Olive oil
   E. Canola oil

8. David Wesson is noted on his extensive work on which of the following oils?
   A. Canola
   B. Cottonseed
   C. Sunflower
   D. Soybean
   E. Both C and D

9. Which oil do we value for its insecticidal properties and use in organic farming?
   A. Castor oil
   B. Soybean oil
   C. Sesame oil
   D. Corn oil
   E. Neem oil

10. Which plant yields the wax used in making naturally fragrant candles?
    A. Copernicia
    B. Elaeis
    C. Myrica
    D. Linum
    E. Aleurites

True/false Questions

1. We can obtain essential fatty acids only from saturated fatty acids.   **True**   **False**

2. Coconut oil is derived from the oily endosperm.   **True**   **False**

3. Ricebran oil has the highest amount of antioxidants of all cooking oils.   **True**   **False**

4. The United States is the leading producer of soybean oil.   **True**   **False**

5. Sunflower oil plants are native to Asia.   **True**   **False**

6. Pure olive oil is considered as of superior quality to virgin olive oil.   **True**   **False**

7. Vegetable oils with low smoking points are suitable for deep frying.   **True**   **False**

8. Corn oil is only a by-product of the corn milling industry.   **True**   **False**

9. Walnut oil is rich in omega-3 fatty acids.   **True**   **False**

10. Nutritionists generally recommend a large consumption of olive oil due to its high percentage of polyunsaturated fatty acids.   **True**   **False**

CHAPTER 8
FLAVOR FOR OUR FOOD

Imagine walking around your school cafeteria. You begin to feel hungry even though it is not even close to your meal time. You can picture that slice of pizza, garlic bread, or cinnamon bun in your mind without even looking at them. You would probably guess other menu items served inside just by the aromas that pass through the air. Thanks goes to some chemical compounds called **essential oils** present in certain plants that impart the characteristic aroma rendered in the food and make you crave them. Such edible aromatic plants are called **spices and herbs**. We use spices and herbs in small quantities to enhance the flavor of the food. Historically, spices and herbs were preservatives and covered the smell of stale meat in the absence of refrigeration facilities. Turmeric is one such spice used in Asian cooking as a marinade and for its anti-microbial activity. Recent research study by Cornell University supports a correlation between climates and spice use. Countries with hotter climates used more spices on meats compared to countries with cooler weather patterns. Many spices may not have any real nutritional value, even though certain spices and herbs are used in herbal medicine. Ginger tea relieves stress headaches, and we know about cinnamon's calming effect.

In Search of Aromatic Plants

While most people tend to think that spices and herbs are terms interchangeably used for the same thing, there are certain differences between spices and herbs. Spices come from different parts of the plants, such as the bark, fruit, roots, stems, leaves, and so forth, whereas most herbs come from the leaves and/or seeds. Most spices originated in tropical parts of the world; herbs came from temperate parts of the world. In addition, spices in general give a very strong aroma while herbs impart only subtle flavors.

Even though spices are plentiful today, in the ancient medieval times of Western civilization, they were precious and in great demand. Spices were much sought after for perfumes, medicine, and flavoring. Adventurers discovered new lands; nations fought wars; and armies conquered countries because of spices.

Records of spice use among ancient people date back to as early as 2500 BCE. These are available in artwork in the pyramid at Giza. The Egyptians used spices in their religious ceremonies, to embalm mummies, and in funeral rituals. Traders had established an active spice trade with southeast Asia and China by 1400 BCE, and the Greeks further expanded it. The Greek physician Hippocrates (400 BCE) listed many spices and herbs, most of which we still use today. The Arabs played the middlemen and controlled the spice routes with caravans crossing thousands of miles of lands. As the Roman Empire grew, the use of aromatic plants became widespread, including everything from lamp oil to perfumes and food. That started a direct trade to India to procure cinnamon, nutmeg, ginger, and cloves in exchange for gold and silver. Historical documents from

65 CE show that a year's worth of cinnamon was burned at the funeral of Nero's wife as a tribute. The spice trade declined during the Dark Ages (until 1096 CE) with the fall of the Roman Empire.

Starting in 1200 CE Europeans explored new passages to China, India, Indonesia, and the Spice Islands (the Moluccas), where the world's most valuable spices existed in plentiful amounts. Marco Polo, a Venetian trader, began his travels to the Orient in 1271 at the age of seventeen. Upon his return after twenty-five years, he became a prisoner of war. During his captivity he wrote the memoirs of his adventures, later published as *The Travels of Marco Polo,* which mentioned the spice plantations in vivid detail. In 1498, Vasco da Gama, a Portuguese explorer, reached Calcutta, India by sailing around Africa's Cape of Good Hope. He established a direct spice trade with India and returned with a variety of Indian spices. Around the same time in 1492,

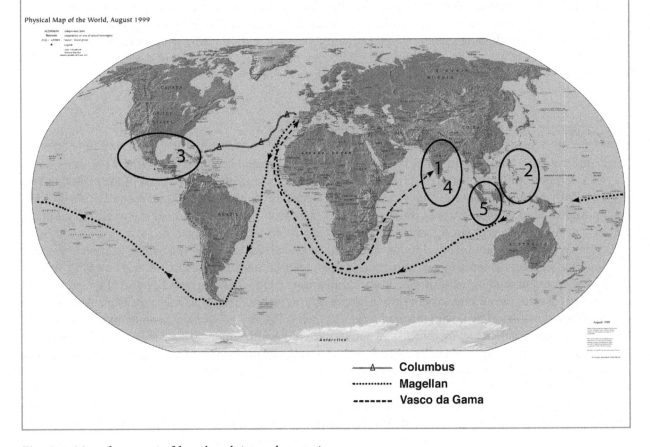

FIVE MOST IMPORTANT SPICES IN THE WORLD

NUMBER	SPICE	NATIVE ORIGIN
1	Cinnamon	Sri Lanka and India
2	Cloves	The Moluccas
3	All spice	Central America and West Indies
4	Pepper, black	Sri Lanka and India
5	Nutmeg	Indonesia

VOYAGES BETWEEN THE FIFTEENTH AND SIXTEENTH CENTURIES

Physical Map of the World, August 1999

⎯△⎯ **Columbus**
············· **Magellan**
------- **Vasco da Gama**

Fig. 8.1 Map of voyages in fifteenth and sixteenth centuries

Christopher Columbus set out in search of Indian spices through his voyage and inadvertently discovered the New World. This eventually introduced vanilla, allspice, and capsicum peppers to the rest of the world. In 1519, Ferdinand Magellan from Spain voyaged on a western route across the Pacific Ocean. Even though Magellan did not return alive, a great deal of black pepper arrived through the remaining crew.

As the popularity for spices grew among common people, competition for the control of the spice trade developed among Portuguese, Dutch, Spanish, and British traders. Wars over the Spice Islands broke out in the fifteenth century and continued for about two hundred years. The Dutch controlled the spice trade until about 1796. The establishment of the British East India Company and the abundant cultivation of spices in French-controlled islands in the Indian Ocean brought an end to Dutch monopoly by 1799.

The United States played a small role in the history of spice trading, beginning its trade in 1672. Captain Jonathan Carnes sailed from Indonesia to Salem, Massachusetts. His trade of black pepper with Asian natives started the first trade center in Salem, which continued successfully over the next ninety years. New contributions from America included chili powders in 1835 and dehydrating techniques for onion and garlic introduced in 1889. World War II brought much popularity over the spices from Europe and Asia, including oregano as a "pizza herb."

Spices no longer remain confined to their original places of origin; a wide variety of aromatic plants grow in Western parts of the world. The United States leads in the procurement of spices today, followed by Germany, Japan, and France. Jamaica grows ginger and allspice extensively.

Spices

Cinnamon

Cinnamon, *Cinnamomum zeylanicum*, is one of the oldest and most valuable spices known. This is one of the exotic spices from the Old World that has generated the interest of spice explorers from the Mediterra-

Fig. 8.2 Cinnamon harvest illustrated in a sixteenth-century French text
Courtesy of the National Library of Medicine.

nean and European regions. It belongs to the laurel family, Lauraceae, which includes plants that give us camphor, avocado, and bay leaves. The species name *zeylanicum* derives from Ceylon, an old name for Sri Lanka, where, along with adjacent parts of India, it is native. Cinnamon is sold whole as "quills" or curls that look like short, brown sticks with rolled margins. These are the products of the inner bark from the trunks and branches of the tree. Cinnamon growers make incisions on the outer bark of the tree to remove a long slice and strip it away. A second incision on this open strip brings out a vertical slice of the inner bark that they cut and roll by hand into *quills.* They allow this bark to ferment chemically and dry, which, when properly sealed, allows it to remain fresh for months. Merchants also sell cinnamon spice in a ground powder form that has a shorter shelf life.

When cinnamon quills appear rough with coarse margins, it may indicate that they are from one of a few other species, *Cinnamomum cassia*, *C. burmannii*, or *C. loureirii*. These plants yield a lower quality cinnamon spice. Workers strip away the entire bark from these plants and then scrape off the outer bark. The process renders a coarse final product that is of lower quality. Quills from *Cinnamomum zeylanicum* have a more delicate and finer flavor than the spices derived from any of the above three species, which we more commonly refer to as "cassia," even though they bear little resemblance to true cassia from Fabaceae. Recent studies indicate that with regular, moderate use cinnamon reduces blood pressure and hypertension in humans.

In Western cooking, consumers use cinnamon to flavor deserts such as applesauce and fruit pies. In Asian cooking they use cinnamon in meat dishes and most commonly as an important ingredient in curry powder. Bay leaves used as a flavoring in meat dishes, soups, and sauces are from the sweet bay, ***Laurus nobilis***.

Cloves

Cloves, *Syzygium aromaticum*, are from Myrtaceae. Members of this family are characterized by plant parts that have this strong flavor imparted by the essential oil, eugenol or clove oil, used in the perfume and soap industry. Cloves are native to the Moluccas (the Spice Islands) and some areas of Indonesia.

Cloves are the unopened flower buds harvested along with the flower stalk when the buds are just turning pink. Processers dry and clean the buds and grade them for export. The "head" of the spice is the flower bud. In less expensive packages of cloves the heads are commonly missing and thus appear empty. Empty cloves lack the superior flavor of the round, flat, unopened cloves. India has used cloves for over two thousand years along with betel leaves and betel nuts that the natives fold into small pockets and chew for calming effects. Today cloves are used as an ingredient in curry powder, meat dishes, and sauces and gravies, as well as a mouth freshener and in toothpaste and medicines. They are also used in cigarettes.

Allspice

Allspice, also known as Jamaica pepper, clove pepper, and pimento, is another member of the same family known by the binomial *Pimenta dioica*. The name allspice refers to its multi-faceted flavor that combines the flavors of cinnamon, cloves, and nutmeg. Allspice is a New World spice and a native of Central America and the West Indies. The spices are actually the dried, whole green berries. The New World, especially Jamaica, still exclusively grows this spice.

Fig. 8.3 Pepper plant
© Jupiter Images, Inc.

Pepper

Black and white peppers are the dried fruits of *Piper nigrum*, Piperaceae. Commercial pepper comes in two commonly sold varieties: black and white.

Spices are the dried drupes that workers pick whole, unripe, and still green growing on a climbing vine. Once picked, they dip the fruits quickly in hot water and then dry them either in direct sun or using a machine. Fermentation occurs during this time along with the rupture of the outer fleshy layers of the fruit, which results in their wrinkled appearance. These are commonly called black pepper or peppercorns. White pepper is from the same plant, except that the growers pick them ripe and allow the fleshy layers of the fruit to decompose and then scrape them off. The white pepper is actually the pit of the drupe. One can use black and white peppers whole or as ground powder in place of cayenne pepper. A rarely seen product in the United States, but common in Asian cooking, are red or green peppercorns that are whole, ripe red or green drupes of pepper pickled in brine or made into freeze-dried products. Pepper is native to India and Sri Lanka, but Viet Nam and Indonesia commonly grow it. The United States is the world's leading consumer of black pepper. Western Kerala in India produces about fifty-two kinds of spices of which they harvest 42,000 tons of pepper each year and export 13,395 tons to the United States. Known as the "**king of spices**," local people view peppers as an antidote for poison and digestive problems as well as a cure for impotence.

Pink peppercorns of the Western cuisine are from a plant unrelated to the genus *Piper* and belong to the genus *Schinus* from Anacardiaceae.

Nutmeg and Mace

Nutmeg and **mace** are two different spices with distinct flavors of their own, but they come from the same fruit of the plant *Myristica fragrans,* Myristicaceae, a native of the islands of Indonesia. There are other species of *Myristica* used as a cheaper alternative to *M. fragrans*. The plants are dioecious. The fruit is a peculiar dry berry that opens up to expose the seed and fleshy, red outgrowths. The seed part is the nutmeg, and the outgrowth of the seed coat, aril, is the mace. The seed is primarily made of endosperm. Growers dry both the seed and the aril before packaging them as spices. Nutmeg has a sweeter flavor and is usually grated fresh while aril imparts a delicate flavor. People use both either whole or ground. They use nutmeg more commonly in desserts, although its use varies among different cuisines, including use in meat dishes and vegetables. The hallucinogenic properties of nutmeg when taken in intermediate doses are well known due to the presence of the chemical compounds myristicin and elemicin. In large quantities it could be toxic. Guatamala and Indonesia are the world's leading producers while Europe, the United States, the Netherlands, India, and Singapore import most of the nutmeg.

Star Anise

Star anise, *Illicium verum,* Illiciaceae, is a fruit from a tree native to southeast Asia. The dried ripe fruits are star-shaped, brown, and thick and used as flavoring and carminative.

Cardamom and Related Plants

Cardamom, ginger, turmeric, and mango ginger are all native to southern Asia and belong to Zingiberaceae. **Cardamom**, *Elettaria cardamomum,* is derived from a triangular pod, called a capsule. Either the small, black, fragrant seeds or the entire pods are used in flavoring the dishes. Traditionally used in dessert items, cardamom has a unique, sweet flavor. One of the leading producers of cardomom is Kerala, India where they call it "Elakka." Used whole or as a crushed spice, they also mix it with curry powder or use it making spiced tea called "Chai" in India. People also used the fruit pods as medicine to treat inflammation, congestion, and as an antidote for poison. Although cardamom used to be unknown in many parts of the West, ice creams flavored with cardamom are now locally available.

Ginger, another member of the same family, is a spice used around the world, especially in Oriental and Jamaican cuisine. The spice comes from the underground modified storage stems of *Zingiber officinale,* called rhizomes. Fresh rhizome imparts a pungent taste due to several nonvolatile compounds, including gingerols. Ginger stimulates hunger hence people use it to cure mild digestive problems. Ginger stewed with tea can relieve stress-related headaches.

When dried, one can store ginger unrefrigerated, but it becomes a poor substitute for fresh ginger. In Western cuisine, cooks use ginger in cookies, soft drinks, candies, and baking. In Asian cuisine, cooks use ginger along with vegetables and meats and in salads.

Turmeric, also known as yellow ginger, known for its yellow dye and cosmetic uses, comes from southeast Asia. Known by the binomial, *Curcuma longa,* it is another member of the ginger family. In India, the natives have used this mildly aromatic spice since 2000 BCE, and it has played an important role in Hindu religion. Religious ceremonies extensively use the ground paste of the fleshy rhizomes or the fine powder of the dry ones. They use them to dye the holy robes used by priests while conducting the rituals and also those worn by brides as they

TABLE 8.1   Important spices around the world

Spice	Binomial	Family	Origin
Allspice	*Pimenta dioica*	Myrtaceae	Central America; West Indies
Cardamom	*Elettaria cardomom*	Zingiberaceae	Southeast Asia
Chili pepper	*Capsicum annuum*	Solanaceae	South America
Cinnamon	*Cinnamomum zeylanicum*	Lauraceae	India; Sri Lanka
Cloves	*Syzygium aromaticum*	Myrtaceae	The Moluccas; Indonesia
Ginger	*Zingiber officinale*	Zingiberaceae	Southeast Asia
Mace	*Myristica fragrans*	Myristicaceae	Indonesia
Mango ginger	*Curcuma amada*	Zingiberaceae	Southeast Asia
Nutmeg	*Myristica fragrans*	Myristicaceae	Indonesia
Pepper	*Piper nigrum*	Piperaceae	India; Sri Lanka
Saffron	*Crocus sativus*	Iridaceae	Eastern Mediterranean
Star anise	*Illicium verum*	Illiciaceae	Southeast Asia
Turmeric	*Curcuma longa*	Zingiberaceae	Southeast Asia
Vanilla	*Vanilla planifolia*	Orchidaceae	Mexico

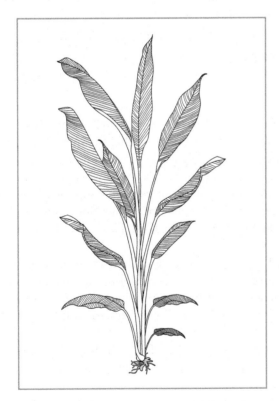

Fig. 8.4 Turmeric plant
Courtesy Rani Vajravelu.

get married. Turmeric is used in south Asian and Middle Eastern cooking as a marinade along with other spice powders. We also know turmeric for its anti-microbial, antiseptic, digestive, and cosmetic uses. The active ingredient is curcumin, which scientists are studying for its effect on various types of cancer and on Alzheimer's disease. Another species of *Curcuma* used in India as a flavoring is **mango ginger**, *C. amada,* which has rhizomes with an aroma of fresh, raw mango. Researchers are conducting extensive studies to test its various medicinal uses, including its cholesterol lowering properties.

Saffron

Saffron comes from autumn crocus, *Crocus sativus* from Iridaceae. Growers pick and dry the orange colored three-part stigmas of this beautiful purple-colored monocot flower. This mild spice is primarily used for its subtle flavor and yellow color. Since they must pick the flowers during a short blooming period and remove the stigmas by hand within a few hours, this spice is one of the most expensive spices known. Approximately one hundred thousand flowers are necessary to get a pound of saffron threads. Autumn crocus is native to the eastern Mediterranean regions, where people have used it since ancient times. India, China, Iran, and Turkey cultivate it widely. Saffron is extensively used in French, Spanish, and Indian cooking. The color and flavor of saffron is due to the chemicals crocin and safranal. The Arabic word zafarn (yellow) contributed to the name of this spice, which imparts a yellow color when cooked with other foods. Iran is the leading producer of saffron. Arabs introduced it in the tenth century.

Vanilla

Vanilla is from *Vanilla planifolia,* a perennial vine from Orchidaceae, and is native to Mexico. The Spanish and Portugese explorers introduced vanilla to Africa and Asia. The word vanilla derives from the Spanish word vainilla, which refers to the fruit pods called capsules. Vanilla plants undergo a pollination mechanism aided by certain bees and tiny humming birds. When plants grow outside of their native origin, growers can accomplish pollination manually by hands. They pick the fruit pods, referred to as "beans," before they become fully mature and place them in the sun for several hours. They wrap such heated pods tightly in blankets and place them overnight to encourage sweating. As they start to turn brown, the slow drying and sweating process continues for a few weeks until the pods turn black and a fine aroma develops. Because of the intensive labor involved, vanilla is second to saffron as the most expensive spice. Pure vanilla extract is made by dissolving out the vanillin from the beans by the ethanol-water percolation method. Manufacturers often make synthetic vanilla from wood pulp or coal tar, which lacks the subtle flavor of pure extract. Recent biotechnology methods attempt to culture vanilla-producing plant cells to reduce the cost of pure vanilla extract. This would interfere with the economy of Madagascar, the leading producer of vanilla beans, which depends heavily on vanilla export for its revenue.

Chili Pepper

Chili peppers (or capsicum peppers) are the fruits of a single genus *Capsicum,* which includes about five cultivated species. The most common one is *Capsicum annuum* included in Solanaceae. This plant is native to South America, but now growers widely cultivate it throughout the tropical parts of the world. When Columbus introduced the capsicum peppers to Spain their use spread throughout Europe and Asia. The fruit is a berry that remains fleshy in bell peppers or eventually dries out as in cayenne peppers. Consumers know of several hundred varieties of chili peppers, which they use as vegetables or spices, either whole or ground into a powder or flakes, or

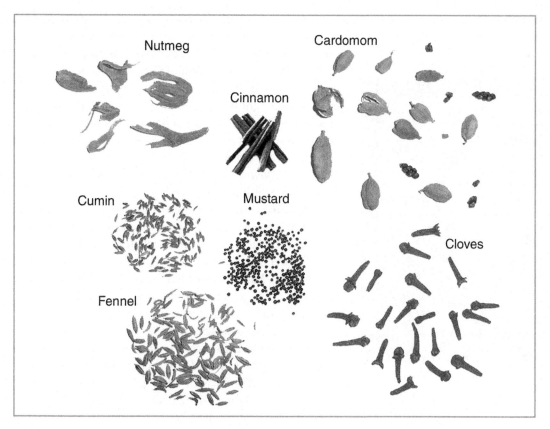

Fig. 8.5 Spices and herbs
Courtesy Rani Vajravelu.

cooked in sauces. Chili powder (or paprika), made by grinding the dried-out fruits, is a main ingredient in Indian spice powders called "masala" or "curry" powder. The burning sensation of this pepper is due to capsaicin, the amount of which may vary depending on the pepper variety. Scientists use the Scoville Heat Units to determine the heat of various peppers, with bell peppers being zero to habanero peppers ranging from 100,000–300,000 units. Jalapeno peppers are within a range of 5,000 units. Peppers are also excellent sources of vitamin C and used in medicine as arthritis cream and as an analgesic. Cayenne pepper stimulates blood circulation.

Herbs

Herbs used in culinary practices derive from the leaves and seeds of certain aromatic plants, such as mint, parsley, basil, dill, and mustard. Most of the common herbs originated in the temperate regions of the world, and people used them for thousands of years.

Cooks can use these herbs fresh or dried, usually in small quantities to enhance the flavor or color of a prepared dish. They can easily grow most herbs on a small land area adjacent to the kitchen.

The majority of herbs are derived from only three angiosperm families, described below.

Carrot Family (Apiaceae)

The distinguishing characteristics of members of the carrot family are their umbrella-shaped inflorescence clusters called umbels and their dry fruits that are schizocarps that split into two mericarps, commonly referred to as the seeds. Besides the vegetables discussed in Chapter 3, this family gives us a multitude of herbs as well.

Parsley, *Petroselinum crispum*, is a biennial herb with bright green dissected leaves. Middle Eastern cooking uses it widely due to its origin in Persia. In Europe and North America, people use parsley as a garnish and also to dispel garlic breath.

Coriander, *Coriandrum sativum*, is the source of two widely used herbs, the soft, leafy cilantro that resembles parsley and the round, dry, brown, nutty-flavored coriander "seeds," which are the schizocarps. Originally from southwest Asia, coriander is one of the popular herbs grown in New World colonies. Middle Eastern, Chinese, African, Indian, and Mexican dishes use cilantro widely. It is best when used fresh and added right before finishing up the dish. Consumers grind coriander seeds (*dhania*) to a powder and use it as a main ingredient in Indian "garam masala" powder, which they use in preparing a saucy, gravy curry.

Dill, *Anethum graveolens*, is an annual herb with finely dissected, needle-like soft leaves.

Ajwain is from *Trachyspermum ammi*, a well-known carminative and aromatic herb native to the eastern Mediterranean and now widely cultivated in India and Iran.

Fennel, *Foeniculum vulgare*, is a similar plant with a harder leaf texture. Many parts of Italy, including Florence, use one of the cultivars of fennel for its fleshy "bulb" as a vegetable. Middle Eastern and Indian cooking uses other varieties for their dry "seeds" (two mericarps of a schizocarp).

Asafoetida is the resin from the stems and roots of a giant fennel, ***Ferula asafetida***, which is native to Iran and west Afghanistan. The brown, dried gum resin sells in hard blocks or in powdered form, which cooks add directly in small quantities either dissolved in water or fried in oil. Its distinctive smell

Fig. 8.6 Cumin
© Ivan Tihelka/Shutterstock.com

is somewhat similar to onion, and Indian vegetarian cooking that uses a lot of lentils favors it as a way to remove flatulence.

Anise, *Pimpinella anisum,* has a licorice flavor, and people use the seeds to flavor candy, liquors, and toothpaste.

Cumin, *Cuminum cyminum,* is another component of Indian curry powders, which cooks use as flavoring in the Netherlands, France, the Middle East, and Chinese and south American countries.

Caraway, *Carum carvi,* is used for its crescent-shaped fruits called "seeds," added to rye breads, liquors, and sauerkraut. Both cumin and caraway are used to cure digestive problems.

Mustard Family (Brassicaceae)

Members of this family are distinguished by the cruciform arrangement of petals, which often form a cross-like pattern. Chapter 3 describes many vegetables derived from members of this family.

TABLE 8.2 Popular herbs around the world

Family	Included Herbs	Binomial
Apiaceae	Ajwain	*Trachyspermum ammi*
	Anise	*Pimpinella anisum*
	Asafoetida	*Ferula asafetida*
	Caraway	*Carum carvi*
	Coriander	*Coriandrum sativum*
	Cumin	*Cuminum cyminum*
	Dill	*Anethum graveolens*
	Fennel	*Foeniculum vulgare*
	Parsley	*Petroselinum crispum*
Asteraceae	Tarragon	*Artemisia dracunculus*
Brassicaceae	Horseradish	*Armoracia rusticana*
	Mustard, black	*Brassica nigra*
	Mustard, brown	*Brassica juncea*
	Mustard, white	*Brassica alba*
	Wasabi	*Wasabia japonica*
Capparaceae	Capers	*Capparis spinosa*
Fabaceae	Fenugreek	*Trigonella foenum-graecum*
Lamiaceae	Basil	*Ocimum basilicum*
	Marjoram	*Origano majorana*
	Oregano	*Origano vulgare*
	Peppermint	*Mentha piperita*
	Rosemary	*Rosemarinus officinalis*
	Sage	*Salvia officinalis*
	Spearmint	*Mentha spicata*
	Thyme	*Thymus vulgaris*
Rutaceae	Indian curry leaf	*Murraya koenigii*

Brassica alba

Fig. 8.7 White mustard
Courtesy Rani Vajravelu.

The genus **Brassica** is known for its seeds derived from dry siliques that give us the black, brown, and white mustards **B. nigra**, **B. juncea**, and **B. alba**, respectively. Greek and Roman times knew of black mustard, which is native to Europe and Africa. Brown mustard, also known as Indian or Chinese mustard, originated from India and parts of China; people often mistake it for black mustard. Several cultivars of this species are known for their thin, basal leaves, either eaten raw or cooked as a vegetable green. The white mustard is European in origin and preferable in British and North American cooking. Hot dog mustard is a mix of two or more kinds with turmeric added for color.

With the popularity of Dijon-style mustard and other specialty blended items, mustard is one of the heavily traded condiments in the world. Within the United States, mustard ranks as the third most popular condiment besides salsa and ketchup.

Before the introduction of black peppers and capsicums, **horseradish**, **Armoracia rusticana**, was a source of hot pungent taste in many parts of Europe. Egyptians have used it since 1500 BCE. In the United States, horseradish is used in Bloody Mary cocktails and in sauces on meats and sandwiches. Unlike other herbs, the large, fleshy taproots of this plant are used to prepare the horseradish sauce. The sharp biting taste is due to sinigrin and Isothiocyanates (also found in wasabi and black mustard). Collinsville, Illinois hosts an annual International Horseradish Festival to boast its extensive cultivation of this condiment.

Wasabi or Japanese horseradish, **Wasabia japonica**, is native to Japan. The Japanese use it in food preparation. Wasabi develops a fleshy root that one should finely grate before use. Because the strong hot flavor dissipates rapidly, merchants usually sell this condiment as a green paste in small tubes. The wasabi that accompanies sushi in the United States is mostly horseradish, mayonnaise, green food coloring, and very little or no real wasabi.

Mint Family (Lamiaceae)

Members of the mint family are readily identifiable with their squarish stem, simple aromatic leaves, and irregular flowers. The Mediterranean region is the center of origin for the mint family even though many included members thrive well in other parts of the world. *Mentha* is the genus for spearmint, *M. spicata,* and peppermint, *M. piperita;* the latter is the most widely grown herb today. In addition to its commercial use in jellies, candies, mouthwashes, toothpaste, ice cream, teas, and medicines, Mediterranean and Indian cooking commonly uses mint leaves in flavoring rice and meat based dishes.

Oregano is the genus for oregano, *O. vulgare,* and marjoram, *O. majorana*. Both are perennial plants. Oregano was introduced to United States after World War II, and its popularity increased as Americans developed a taste for Italian food, especially pizza. Other well-known herbs in the same family are thyme, basil, rosemary, and sage.

Thyme, *Thymus vulgaris,* is a versatile herb that one can use in stews, soups, meats, and so forth. People use thyme oil for relieving cold symptoms, depression, and muscular pain. They also know it for its antiseptic and fungicidal properties.

Basil, *Ocimum sanctum,* has a mild, sweet aroma and holds a sacred role in Hindu religion. Referred to as "Tulsi" in India, the people consider it the symbol of purity, and Hindus worship it every day. It is valued for its healing power for the common cold and fever as well as for treating skin and eye disorders. Ayurvedic medicine refers to its use. Among many varieties of basil, the sweet basil used in Italian cooking and the holy basil used in Asian cooking are the most common. Many chefs around the world consider basil the "king of herbs." Classic pesto sauce is a mix of sweet basil and olive oil, among other ingredients. Basil is best when used fresh and added right before the completion of a cooked dish.

Rosemary, *Rosemarinus officinalis,* is a woody, fragrant shrub with needle-like leaves that is native to the Mediterranean region. The fragrant leaves are used in herbal tea for calming effects and the oil in aromatherapy, although it is not safe to consume internally. European herbal medicine values rosemary, and people have used it to treat arthritis, headache, digestive problems, eczema, bruises, circulatory problems, and yeast infection.

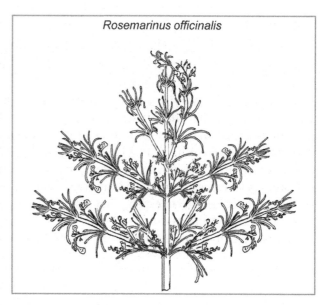

Rosemarinus officinalis

Fig. 8.8 Rosemary
Courtesy Rani Vajravelu.

Sage, *Salvia officinalis,* is a small, woody plant with a slight peppery flavor. Many parts of Europe commonly use it, including Germany, France, Italy, and Britain, to flavor poultry, pork, mutton, and especially stuffing and sausages. Sage was the herbal remedy most commonly used in olden days. When taken in the recommended doses, consumers have known sage to be effective in the treatment of indigestion, depression, and hypertension, as well as for insect bites and skin infections when applied externally. Growers raise a number of cultivars of this plant mainly for ornamental purposes rather than for herbal uses.

Miscellaneous Herbs

Capers, *Capparis spinosa,* Capparidaceae, is a flavor enhancer in Italian, French, and Greek salads, dressings, sauces, and a variety of main dishes. The immature flower buds or the ripe berries of this spiny plant bush native to the Mediterranean region are usually packed with coarse salt and vinegar. Ancient Greek and Indian Ayurvedic medicine also used capers to treat various ailments, including flatulence, dropsy, and rheumatism. The bioflavonoid rutin is a valued antioxidant.

Tarragon, *Artemisia dracunculus,* is from Asteraceae. The plant is native to Eurasia, but it extends to Asia and to western parts of North America. French cooks extensively use the dried, mildly aromatic, bittersweet leaves of this herb, particularly with chicken and fish. In Armenia, Russia, and Persia, they make carbonated soft drinks from sugary tarragon syrup. Tarragon became introduced to the United States only in the nineteenth century, and people use it mostly as tarragon vinegar.

Indian curry leaf, *Murraya koenigii,* is native to India and a member of the Citrus family, Rutaceae. East Asian cultures use the fresh leaves as a flavoring either fresh or fried with popped mustard seeds in oil and added on top of a cooked dish.

Fenugreek, *Trigonella foenum-graecum,* is one of the oldest medicinal herbs and a member of Fabaceae. Researchers believe it originated in the Mediterranean region. Indian, Ethiopian, and Middle Eastern cooks use the small, stony, angular, yellow-brown seeds. In India, they also cook the young leaves as greens and use the dried ones for flavoring (methi leaves). People have used fenugreek widely as a digestive aid and as a galactagogue (to increase milk production). Modern health food stores sell them in capsule form.

Curry Powder

A description of spices and herbs cannot be complete without some information on curry powder. The word "curry" originated in south India, where cooks determine the mixed proportions of different spices to prepare a spicy dish of cooked vegetables or meats. Indian cooks have access to a variety of spices and often prepare them fresh in widely varying compositions to make an array of curry flavors. Curry powder is a modern, convenient variation where the dried spices are ground ahead of time in a fixed proportion and stored in airtight containers. Based on the desired taste and the main ingredient used in a cooked dish, people grind curry powders individually for fish, poultry, lamb, or vegetables. Most standard recipes for Indian curry powder include coriander, chili pepper, turmeric, cumin, fenugreek, and asafoetida. Cooks add other ingredients, such as cinnamon, cloves, cardamom, mace, nutmeg, and black pepper, if the dish involves a meat or if they desire a spicier version of a vegetable dish. They also frequently add ginger, garlic, and onion to dishes to enhance the flavor and health benefits.

Please refer to the concoction of spices and herbs on display in the attached CD.

Just for Fun

A simple recipe for homemade curry powder:

Dry red chilies: 2; coriander seeds, whole cloves: 2 tablespoons each; cinnamon quill; cardamom: 2 pieces each; 1 whole nutmeg; 1-inch piece of turmeric.

Use a clean coffee grinder and grind all dry ingredients together.

Cook with rice or any of your favorite meat or vegetables. A standard amount is 1/2 teaspoon of powder for up to four servings.

Review Questions

Give brief answers in the given space.

1. List all of the spices discovered in the Old World tropics.

2. Explain the processing involved in the preparation of cinnamon quills.

3. Describe how one could derive white and black pepper from the same plant.

4. Which spice could be hallucinogenic, medicinal, and also toxic? And why?

5. List all the spices derived from Zingiberaceae.

6. Explain the reasons behind the high prices involved with saffron.

7. Distinguish between "spice" and "herb."

8. List the families with the multitude of herbs.

9. Explain why "sushi wasabi" served in American deli may not have wasabi in it at all.

10. What are the main ingredients in a basic curry powder?

Multiple Choice Questions

For each question, choose one best answer.

1. Cinnamon is native to _____.
    A. an island in the Atlantic Ocean
    B. New Guinea
    C. Jamaica
    D. Sri Lanka
    E. Mexico

2. A spice, which has the combined flavor of cinnamon, nutmeg, and cloves, is:
    A. *Zingiber officinale*
    B. *Crocus sativus*
    C. *Pimenta dioica*
    D. *Elettaria cardomomum*
    E. *Myristica fragrans*

3. Which of the following is the "King of spices"?
    A. Allspice
    B. Jamaican pepper
    C. Garlic
    D. Black pepper
    E. Capsicum pepper

4. Which of the following spices do Hindu religious rituals use extensively?
    A. Cardamom
    B. Saffron
    C. Cinnamon
    D. Turmeric
    E. Ginger

5. Which of the following pairs of spices could come from the same fruit?
    A. Cloves and cinnamon
    B. Cardamom and black pepper
    C. Black and white mustard
    D. Nutmeg and mace
    E. Both C and D

Match the following chemical compounds from Column 1 to the spice/herb name listed in Column 2. You may use a choice once, more than once, or not at all.

Column 1

6. Eugenol

7. Myristicin

Column 2

A. Nutmeg

B. Chili pepper

C. Horseradish

D. Black pepper

E. Cloves

8. What two spices are from modified, underground storage rhizomes?
   A. Cloves and cardamom
   B. Ginger and turmeric
   C. Horseradish and ginger
   D. Ginger and garlic
   E. Onion and horseradish

9. In 1271, _____ began his travels at seventeen years of age to the Orient and witnessed the spice plantations.
   A. Theophrastus
   B. Marco Polo
   C. Christopher Columbus
   D. Ferdinand Magellan
   E. Vasco da Gama

10. Which of the following spices has uses in desserts, medicines, toothpastes, and also in cigarettes?
    A. Cinnamon
    B. Cloves
    C. Black pepper
    D. Nutmeg
    E. Bell pepper

True/false Questions

1. Ginger is native to Jamaica.   **True**   **False**

2. Chinese mustard is also consumed as a leafy green.   **True**   **False**

3. Jalapeno peppers are more pungent than habanero peppers.   **True**   **False**

4. New biotechnology methods attempt to reduce the price of pure vanilla.   **True**   **False**

5. Commercially sold coriander and cumin seeds are actually not real seeds.   **True**   **False**

6. A year's worth of cloves was burnt as a tribute to Nero's wife.   **True**   **False**

7. Asafoetida is the dried root of a giant fennel.   **True**   **False**

8. Mustard, capers, and tarragon are all members of the mustard family.   **True**   **False**

9. The United States spice trade started with black pepper.   **True**   **False**

10. Traditional Indian curry is a mix of different spices in varying proportions. **True**   **False**

CHAPTER 9
PLANTS OF MEDICINAL VALUE

Some plants, more than others, synthesize chemical substances useful to the health of humans and animals. These substances include alkaloids, glycosides, phenols, tannins, essential oils, gums, and resins. We also know many of these as secondary metabolites as opposed to the primary metabolites that include carbohydrates, proteins and fats. The function of secondary plant products varies between the plant species and mainly includes attracting prospective pollinators and protection from predators and nearby competing species. People from prehistoric times have understood the therapeutic effects of these secondary plant products on human health. Indigenous people of all continents have relied on local plants for the treatment of illness through the knowledge passed on from their predecessors. They mostly kept herbal healing methods secret and transferred them from generation to generation by word of mouth—from father to son, from mother to daughter, and from the tribal leader to a loyal follower.

History of Herbal Medicine

Archaeological studies suggest that medicinal plants were in use among Neanderthals about sixty thousand years ago. Cave paintings in France dated between 25,000 and 13,000 BCE illustrate the use of certain plants as healing agents. Sumerian clay tablets from 2500 BCE describe the use of opium poppy, thyme, and laurel in medicine. About the same time, a Chinese herbal book mentions the use of *Ephedra*. The Ebers Papyrus, an Egyptian written record from 1550 BCE, mentions the use of mandrake and garlic for the treatment of pain and heart disorders respectively. The Rig-Veda, one of Hinduism's sacred records, talks about the use of snakeroot for sedative effects still in use today. The medical manuscript of the Aztec tribes created in 1592 BCE provides an illustrated record of traditional medicinal plants.

Ancient Greeks and Romans made use of the medicinal plants as did other areas of the world. Writings preserved some of these practices. One such notable work is that of a Greek physician, Hippocrates (460–377 BCE) who believed that bodily problems caused sickness and used herbal remedies in his treatments. Since his work and that of his followers set the basis for the study of modern medicine called pharmacology, we know Hippocrates as the **Father of Medicine**.

Theophrastus (271–287 BCE), a student of Aristotle, made similar contributions to the study of medicinal plants through his botanical work, **Historia Plantarum**. It provided detailed descriptions of plants, including the ones from the Gardens of Athens.

A significant contribution in the first century CE came from Pedanius Dioscorides (40–90 CE) of Greek origin who wrote a five-volume set called **De Materia Medica** (The material for medicine) that was a compendium of more than a thousand simple plant drugs. It remained an authoritative reference until the 1600s.

During the Middle Ages that followed, also referred to as the Dark Ages, contribution to medicinal plant knowledge was minimal, although a link between botany and medicine developed more closely. Local monasteries provided plants for common disorders; people predominantly used folk medicine. Starting in the eleventh century, medical universities began to appear, and herbal knowledge from the east started to spread to most of the Arab world. The entire Middle East had a rich history of herbal medicine. There are medical texts from the ancient cultures of Mesopotamia and Egypt that describe and illustrate the use of many medicinal plant products. Some of the plants mentioned were castor oil, linseed oil, and white poppies. With the invention of printing presses, hundreds of herbals, including Historia Plantarum and De Materia Medica, were published.

The great age of herbals began in the fifteenth century. One important contribution was by Paracelsus (1493–1541), raised in Switzerland. He believed that he had superior insights into the prevention of illness and its cure due to his extensive knowledge gained through travel in much of Europe and his doctoral degree. He publicly spoke against traditional medicine by burning the medical books, including those of Theophrastus and Dioscorides. According to Paracelsus, God placed plants on earth for human use. Such plants, according to him, exhibit distinct "signatures"—that is, morphological resemblances to certain human body parts that indicate their uses. For example, the walnut with its convoluted appearance resembles the human brain and thus can cure ailments related to the brain. His idea, called the **doctrine of signatures**, received great recognition at that time, but it soon became unpopular.

The experimental approach to medicine began in the seventeenth century. **William Withering** (1741–1799) discovered the medicinal properties of foxglove plant to treat heart problems. Through the first half of the twentieth century, medical botany improved tremendously as researchers isolated modern medicines from purified plant extracts or as they made synthetic drugs based on natural products. Some important drugs of this time were morphine from the opium poppy and synthetic salicylic acid based on the active ingredients in willow plants. The direct use of plants and plant extracts started to decline in the United States compared to the rest of the world beginning in the late twentieth century. Although other countries are conducting substantial research on medicinal plants, researchers have scientifically studied only a very few of the healing herbs that exist among the angiosperm plants. Over 25% of the prescription drugs in the United States derive from plant ingredients, and even more synthetic drugs have a basis in active principles in plants. For example, modern drugs for cancer treatments and chemotherapy were derived from research on traditional plant remedies. Recently a renewed interest in investigating the medicinal uses of plants is developing with the help of some leading pharmaceutical and research institutions. One such is an attempt to gather folk knowledge of medicinal plants from the indigenous people of rainforests and other tribes. About 75% of the plant-derived pharmaceutical medicines directly correlate with medicinal uses by native tribes. Incorporating traditional medicine into the healthcare system is one of the responsibilities undertaken by the World Health Organization (WHO), established in 1948. WHO estimates that 80% of the world population currently uses herbal medicine for some aspect of primary health care. Herbal medicine is a common element in India's Ayurvedic system, in Chinese oriental medicine, in homeopathy, and in America's shamanism.

The Phytochemicals

There are two major classes of secondary plant products used in medicine: alkaloids and glycosides.

Alkaloids

Alkaloids are basic, contain N_2, taste bitter, and have dramatic effects on the nervous system. About 50,000 alkaloids are found in numerous plant families. Scientists classify them based on their effects on the human body. Caffeine is a stimulant; nicotine and cocaine are addictive; morphine and codeine are painkillers; quinine and ephedrine are therapeutic; atropine is hallucinogenic; strychnine is toxic. Some families that are rich in alkaloids are Fabaceae (the legume family), Solanaceae (the nightshade family), and Rubiaceae (the madder family). It is often the dosage that determines if an alkaloid is medicinal, hallucinogenic, or toxic. The presence of tropane alkaloids characterizes some members of Solanaceae, which share similar chemical structures and produce similar physiological effects when consumed. They dilate the pupils of the eyes and blood vessels, lessen pain, and activate failed hearts following cardiac arrest. Because of this medical practitioners use them in small amounts to ease nausea. In high doses they are extremely toxic. *Atropa* and *Hyoscyamus* are rich in tropane alkaloids.

Glycosides

Glycosides consists of a sugar molecule called glycone, attached to a non-sugar active component called aglycone. Glycosides are bitter, have a marked physiological effect on animals, and are inert until water or enzymes activate them. Many plants store glycosides in an inactive form. Such plant glycosides have an

TABLE 9.1   Phytochemicals used in medicine

Active Principle	Plant Source	Plant Binomial	Family	Uses
Aloin	Burn plant	*Aloe vera*	Liliaceae	Skin related conditions
Artemisinin	Wormwood	*Artemisia annua*	Asteraceae	Malaria
Atropine	Belladonna	*Atropa belladonna*	Solanaceae	Ophthalmology and more
Cannabinoids	Marijuana	*Cannabis sativa*	Cannabinaceae	Glaucoma; nausea in chemotherapy
Chymopapain	Papaya	*Carica papaya*	Caricaceae	Back pain
Cocaine	Coca	*Erythroxylon coca*	Erythroxylaceae	Anesthesia
Colchicine	Meadow saffron	*Colchicum autumnale*	Liliaceae	Gout; potential drug for cancer
Digitoxin	Foxglove	*Digitalis purpurea*	Scrophulariaceae	Heart failure
Ephedrine	Ephedra	*Ephedra spp.*	Ephedraceae	Asthma; allergy
Henbane	Henbane	*Hyoscyamus niger*	Solanaceae	Pre-anesthetic medication
Morphine; Codeine	Opium poppy	*Papaver somniferum*	Papaveraceae	Pain killer; cough suppressant
Podophyllin	Mayapple	*Podophyllum peltatum*	Berberidaceae	Testicular cancer
Reserpine	Snakeroot	*Rauwolfia serpentina*	Apocynaceae	Hypertension; mental illness
Quinine	Fever bark tree	*Cinchona succirubra*	Rubiaceae	Malaria
Salicin	White Willow	*Salix alba*	Salicaceae	Fever; ache; inflammation
Taxol	Pacific yew	*Taxus brevifolia*	Taxodiaceae	Ovarian and breast cancer
Vinblastine; Vincristine	Periwinkle	*Catharanthus roseus*	Apocynaceae	Leukemia; lymphoma

important role in medicine. Scientists classify glycosides based on the aglycone molecule that can range from cyanide, steroidal, alcohol, phenol, flavonoid, saponin, and sulfur, as well as many other less commonly known groups.

Cyanogenic glycosides exist in cassava plants and in the pits of some Rosaceae stone fruits like cherries, almonds, and peaches. The chemical compounds in the fruit pits release only when an herbivore bites them. Steroidal glycosides have stimulating effects on heart muscles and occur in foxglove. Salicilin is an alcoholic glycoside; tannin, phytoestrogens, and anthocyanins are phenolic glycosides. There is a large group of flavonoid glycosides, including quercetin and hesperitin, which have an antioxidant effect. Saponin glycosides have expectorant and contraceptive effects; sulfur glycosides, also known as thioglycosides like sinigrin, occur in horseradish.

Plants Used in Traditional and Modern Medicine

Cinchona

Cinchona succirubra, the quina-quina, is one of many species that gave us the medicine for the treatment of malaria. Even in this century, malaria is still the world's most prevalent disease. Two to three million people die each year from malaria. Some of the common symptoms of malaria are fever, chills, anemia, and enlarged spleen. In extreme cases, it leads to cerebral malaria characterized by convulsions and coma, which can lead to death. Ancient Egyptian records from 3,500 years ago show the occurrence of malaria.

Female mosquitoes infested with protozoan parasites cause malaria, even though in early days people mistakenly believed that contaminated "bad air" caused this disease. An English physician, Sir Ronald Ross, studied the details of malaria and suggested the mode of transmission was through mosquitoes. He received a Nobel Prize in 1902 for his remarkable work on malaria.

Cinchona is native to the Andes Mountains in South America. The Incas of Peru were aware of the effects of bark juice to treat the fever. Obtaining the knowledge from the Incas, the Jesuit missionaries tried the bark juice to treat malaria. One of the treated individuals who miraculously survived malaria was the Countess of Cinchon, hence the name Cinchona to the genus name, adopted later on by Linnaeus. In 1820, laboratory synthesis of the most active compound, quinine, replaced the bark extraction. **Quinine**, though effective in treating malaria, left patients with many side effects, such as, blurred vision, tinnitus, depression, and miscarriage. The medical profession is using **chloroquine**, a synthetic drug that is less toxic, more effective, and has fewer side effects. Research for more natural sources is still going on around the world. One promising source is wormwood, *Artemesia annua*, known in China for treating fever since 168 BCE. This plant contains artemisinin, a terpene that shows fewer side effects than the other two drugs.

Rauwolfia

Rauwolfia serpentina, *the snakeroot*, is a small, woody perennial. In India, where it is native, traditional Hindu healers have used this plant for the treatment of "moon disease" or lunacy and as an antidote for snakebites. Its Sanskrit-derived name *sarpagantha* has been in use for over four thousand years and mentioned in the Vedas of India. The coiled roots of this plant resemble a snake, a perfect example for the Doctrine of Signatures. Chewing the snakeroots for their calming effects was a practice among the sages of India. Roots are used for treating hypertension and to reduce blood pressure, and administered to patients with schizophrenia. Due to the side effects that lead to depression and tremors it is no longer used to treat hypertension but used in the treatment of mental illness in the form of an alkaloid called **reserpine**, which scientists isolated from the roots in 1952 in Switzerland. This drug altered the conventional practices in medical institutions and now is administered as

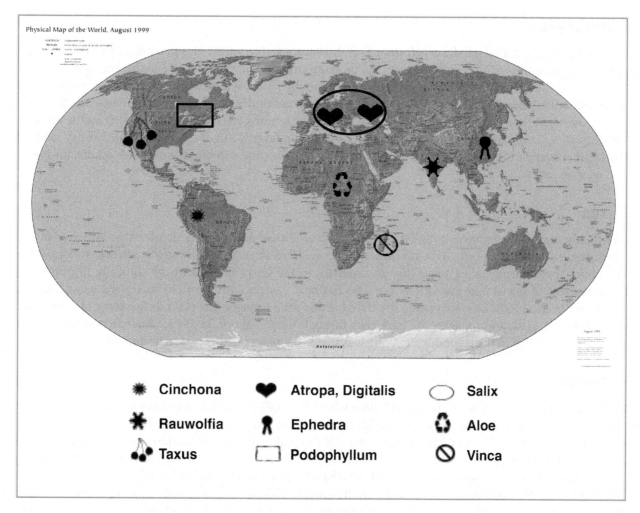

Fig. 9.1 Map of most important medicinal plants and their origin

an alternative to electric shock for certain types of mental illness. Natural extracts of this plant are less expensive compared to synthetics. **R. tetraphylla** is also a source of alkaloids. Most of the world's supply comes from *R. serpentina* located in India, Pakistan, and Java.

Catharanthus

Catharanthus roseus, also known as ***Vinca rosea***, is the rosy periwinkle, a perennial herb from the Madagascar islands. Traditionally healers have used this plant to treat diabetes. In 1957, researchers isolated and marketed two important alkaloids, **vinblastine** and **vincristine**, both extracted from the leaves. Physicians use these extensively in the treatment of certain forms of leukemia and lymphoma.

Colchicum

Colchicum autumnale, the meadow saffron, is a monocot plant being investigated for its potential use as an anti-cancer drug. The alkaloid, **colchicine**, is extremely poisonous and causes burning of the mouth and throat, fever, vomiting, abdominal pain, and kidney failure. Despite its toxicity, it has approval for treatment of gout and is also used in plant breeding firms to promote polyploidy.

Podophyllum

American mayapple, **Podophyllum peltatum**, is an herbaceous, rhizomataceous perennial plant that contains anti-tumor alkaloids mainly extracted from the underground rhizomes. It is native to the forests of southern Canada and the eastern United States. Healers traditionally have used this plant for the treatment of skin disorders and tumorous growths. Every part of the plant except ripe fruits is poisonous and may result in salivation, diarrhea, headache, fever, coma, and other drastic symptoms that may vary depending on the ingested quantity. It contains toxic compounds called podophyllin. In large doses, it can cause death. Today medical field uses mayapple alkaloids as drugs to treat testicular tumors and combined with other agents, for breast and ovarian cancer.

Taxus

Another source of anti-cancer drugs is the Pacific yew, **Taxus brevifolia**, a conifer that is native to the northwestern coastal regions of North America. A medium-sized tree with small needles, it contains a chemotherapeutic drug called **taxol** that accumulates in the bark. Scientists have clinically proven it for the treatment of breast and ovarian cancer. The National Cancer Institute first discovered its anti-tumor properties in the 1960s. Overharvesting the trees for this drug brought this plant to the verge of extinction. Attempts were made to propagate the bark cells through tissue culture. In 1994, researchers synthesized a semi-synthetic taxol in the laboratory, making it available for cancer treatments in unlimited amounts.

Ephedra

Ephedra sinica is known as ma-huang in China where it is native. The Chinese have known about it since 3000 BCE for its traditional use on asthma, cold, and hay fever. It is another conifer plant but mentioned here due to its important role in medicine. Other species of this genus common in western Asia and America have local uses in medicine as well. Mormon tea prepared by Native Americans and Mormons is made from brewing the plants in water. Researchers identified the alkaloid **ephedrine** in 1887, which has sold as a drug since the 1920s. Due to their decongestant properties, the medical profession uses ephedrine and pseudoephedrine for treating bronchitis, asthma, and allergy. Due to a high rate of serious side effects leading to death at times, the Food and Drug Administration (FDA) banned products containing ephedrine and pseudo-ephedrine as over-the-counter medications in 2006.

Atropa

Atropa belladonna is the deadly nightshade plant related to tobacco, the tomato, and the white potato. It is a perennial herb and one of the important species of the nightshade family with extremely toxic or hallucinogenic tropane alkaloids. The name derives from the Greek god Atropos (=inflexible), from the time of Theophrastus. **Atropa** is native to Europe, western central Asia, and some parts of northwestern Africa, but it exists in a naturalized state in many other parts of the world. Historically, Italian women used the leaf juice of belladonna ("beautiful lady" in Italian) as eye drops to obtain dilated pupils that gave a dreamy, intoxicated look believed fashionable at that time. The medical profession uses **atropine** in optometrics and ophthalmology for pupil dilation and to cure asthma, irritable bowel syndrome, gastric ulcers, hay fever, and whooping cough, as well as to reduce tremors in patients with Parkinson's disease.

Hyoscyamus

Hyoscyamus niger, the stinking nightshade, has historical usage in combination with other solanaceous plants with tropane alkaloids as a preanesthetic drug. **Hyoscyamine** and **scopolamine** are abundant in the foliage and seeds of this plant, also referred to as henbane.

Digitalis

Digitalis purpurea, the foxglove, is a rich source of cardioactive and steroidal glycosides concentrated in the entire plant, especially the leaves. Even a small amount can bring a severe headache, hallucinations, and deadly effects on the heart. **Digitoxin** is a cardioactive glycoside extracted from the seeds or leaves that has medicinal use in the case of heart failure. The drug strengthens heart muscle, slows the heart rate, and helps with fluid retention. Medical literature has described *digitalis* since the late eighteenth century. William Withering studied its clinical use and used it to treat dropsy. Dropsy is a condition characterized by bloating due

Fig. 9.2a Aloe
© Nevada31/Shutterstock.com

Fig. 9.2b Foxglove
© Moolkum/Shutterstock.com

Fig. 9.2c Snake root
Courtesy Rani Vajravelu.

Fig. 9.2d Rosy periwinkle
Courtesy Rani Vajravelu.

to fluid accumulation, nausea, excessive sweating, and rapid heartbeat. Today we know it as congestive heart failure. In 1785 Withering published a book titled *An Account of the Foxglove and Some of Its Medicinal Uses: With Practical Remarks on Dropsy and Other Diseases.* Several million heart patients in the United States rely on *digitalis* for their primary treatment.

Foxglove grows wild in the open fields and along the riverbanks and roadsides throughout Europe, the western part of central Asia, and in northwest Africa. The folklore of England associates foxglove with witches and fairies. The beautiful purple-colored, funnel-shaped flowers contribute to stories that fairies wore them as hats or petticoats and gave them to foxes to wear as gloves to help them catch chickens. It was considered bad luck bringing the plants indoors. They had an important role in traditional medicine, including treatment for dropsy.

Salix

Salix alba, the white willow, is a source of **salicin**, a glycoside of salicylic acid. Growers extract it mostly from the inner bark of the tree. White willow is native to Europe and western parts of central Asia. Since the time of Dioscorides, local people in Egypt, Greece, and many parts of Europe and America have used the leaves and bark juice of white willow and other related species for the treatment of insomnia and minor aches. In 1753, Reverend Edmund Stone in England studied its medicinal properties for the treatment of fever and chills. Since laboratory synthesis began in the mid 1800s, the use of salicylic acid spread rapidly for the treatment of rheumatic fever, gout, rheumatism, pain, and inflammation. Since it caused gastric distress in patients when taken internally, today people use it topically for various skin ailments, including warts.

Spirea

Felix Hoffman discovered **aspirin** in 1898 as an alternative to salicylic acid from another plant source, *Spirea*, a member of the rose family. Aspirin derives from "a" for acetylsalicylic acid, and "spirin" from *Spirea*, the plant source for salicylic acid. This gave rise to an important class of drugs known as non-steroidal anti-inflammatory drugs (NSAIDs). Aspirin's value is for its three classic properties: as an anti-inflammatory, as antipyretic (fever-reducing), and as analgesic (pain-relieving). According to a recent survey, Americans take an average of fifty million aspirin tablets a day.

Aloe

Aloe vera is a short-stemmed, semi-tropical plant native to Africa but widely cultivated throughout the world. This is a monocot plant with a dense whorl of thick, succulent, lance-shaped leaves with spiny margins. A few varieties of this species have smaller leaves with white speckles on the surface. Thick clear gel and mucilage fill the fibrous green leaf epidermis, both of which people take internally or use externally as a moisturizer in herbal medicine. Aloe has been in use since 1500 BCE and mentioned in Egyptian papyrus. According to historical documents, Alexander the Great conquered islands in the Indian Ocean to get a constant supply of *aloe* that naturally grew over there. Japanese, Chinese, Indian, Greek, and Native American folk medicine have long known the use of *aloe*. The leaves also have a sticky latex packed with glycosides **aloin**. Medical practitioners have proven aloin very effective and use it in the treatment of skin disorders, burns, cuts, and eczema, as well as to ease pain and reduce inflammation. Due to its reputation for the treatment of burns, we know it as "the burn plant." Many cosmetic products, including soaps, shampoo, lotions, make–up, and skin conditioners contain aloe extracts. Practitioners often use aloin with belladonna extracts to treat intestinal ailments.

Carica

Carica papaya is the source of latex that yields some enzymes used in the treatment of back pain. Invasive surgery to replace a misplaced disk was the only option for many patients who suffered from back pain caused by a herniated disk. Doctors carefully administer chymopapain, a protein degrading enzyme, on a large part of the disk to dissolve the noncollagen part of the area. This relieves the pressure on the adjacent nerves and reduces the pain. The procedure became introduced in the mid-1960s, and the FDA approved its use in 1983. The drawback to this surgery is the allergic reactions developed by some patients to the enzymes of papaya.

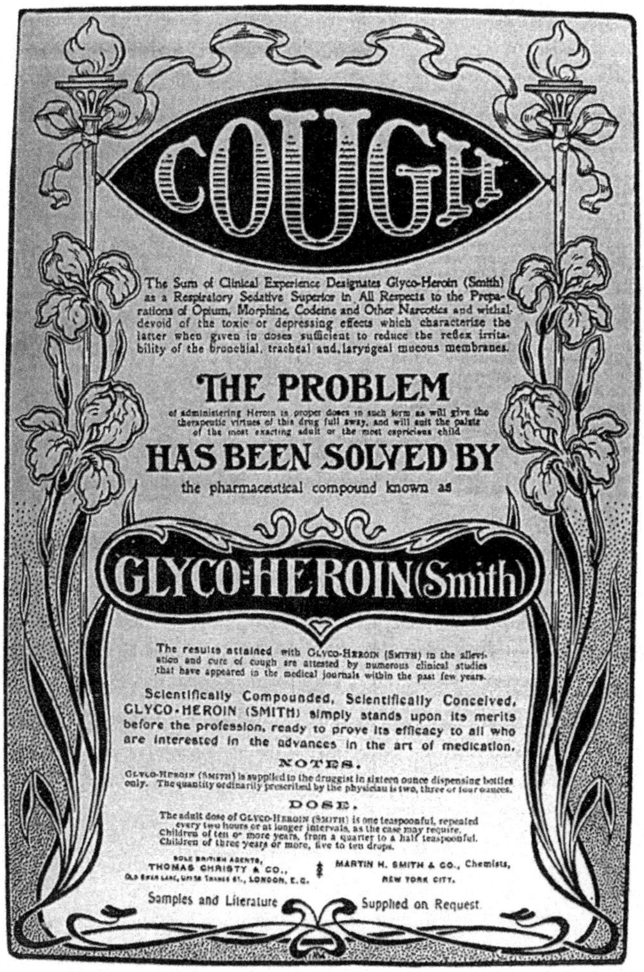

Other Plants of Medicinal Value

Medical practitioners traditionally used marijuana, the opium poppy, and coca plants as pain killers. They altered and improved administration procedures after the discovery of their addictive properties. We will briefly discuss the medicinal values of these three plants.

Cannabis sativa, marijuana, contains terpenophenolic compounds known as cannabinoids. The medically useful compound is tetrahydrocannabinol (THC), isolated in 1965. For the past twenty years or more, doctors have used THC as a pain reliever, to reduce the pressure in the eyes of glaucoma patients, and as an anti-nausea agent for chemotherapy patients. Due to concerns about addictive properties, scientists are studying synthetic variants of THC.

Papaver somniferum, the opium poppy, is a source of about twenty-four different

Fig. 9.3 An advertisement from the beginning of the nineteenth century showing heroin's use in medications. Its addictive properties were discovered later.

From *Plants and Society*, *5e* by Estele Levetin & Karen McMahon, 2008. Reproduced by permission of The McGraw-Hill Companies.

alkaloids, of which medical practitioners have used morphine as a potential pain killer and as an anesthetic for surgical procedures. Researchers introduced a semi-synthetic heroin as an analgesic and cough suppressant. Due to their severe addictive properties, the United States prohibits the use of these drugs except under strict medical supervision. Codeine is a milder, non-habit-forming alkaloid used in over-the-counter pain relief medications and many prescription drugs. Papaverine's use in drugs is to treat intestinal spasms and diarrhea.

Erythroxylum coca, the coca plant, is the source of alkaloid cocaine. The law restricted the medical use of this alkaloid to producing local anesthesia. In the United States, cocaine was sold as over-the-counter medicine until 1916. In Europe, its use is sometimes as a surface anesthetic to numb local areas before surgery. The United States has not used it as medicine.

Name: _____   Date: _____

Review Questions

Give brief answers in the given space.

1. Explain the doctrine of signatures with at least two examples.

2. What is the role of WHO in modern medicine?

3. List at least five alkaloids from five different plant sources.

4. Explain how the chemical nature of glycosides makes them suitable for medicinal use.

5. Which two medicinally important plants are gymnosperms, and what do we use them for?

6. Explain William Withering's contribution to medicine.

7. Briefly narrate the history behind the discovery of aspirin.

8. Debate the pros and cons of chymopapain in medicine.

9. Even though snakeroot had many traditional cures, there are restrictions on its modern medical use. Why?

10. List all the glycosides derived from medicinally useful plants.

Multiple Choice Questions

For each question, choose one best answer.

1. *De Materia Medica* was written by:
    A. Paracelsus
    B. Theophrastus
    C. Aristotle
    D. Dioscorides
    E. Withering

2. The following belong to a class of chemical compounds. Pick out the exception.
    A. Atropine
    B. Morphine
    C. Digitalin
    D. Quinine
    E. Cocaine

3. Reverend Edmund Stone worked on the medicinal properties of:
    A. Foxglove
    B. Willow
    C. Aloe
    D. Cinchona
    E. None of the above

4. We obtain reserpine from the _____ of _____.
    A. rhizome; Podophyllum
    B. leaves; Atropa
    C. roots; Rauwolfia
    D. leaves; Ephedra
    E. bark; Cinchona

Match the chemicals in Column 1 to the plant they derive from in Column 2. You may use an answer once, more than once, or not at all.

Column 1   **Column 2**

5. Quinine   A. snakeroot

                                          B. pacific yew
6. Vincristine
                                          C. fever bark tree
7. Taxol   D. periwinkle

                                          E. willow

8. A plant family high in tropane alkaloids would be:
    A. Fabaceae
    B. Papaveraceae
    C. Solanaceae
    D. Rubiaceae
    E. Apocynaceae

9. Which of the following pairs of plants belong to the same family?
    A. opium poppy and marijuana
    B. periwinkle and snakeroot
    C. belladonna and periwinkle
    D. fever bark tree and pacific yew
    E. mayapple and foxglove

10. Ancient Italian actresses used leaf extracts of _____ to enhance the beauty of their eyes.
    A. *Cathranthus roseus*
    B. *Digitalis purpurea*
    C. *Atropa belladonna*
    D. *Salix alba*
    E. *Aloe vera*

True/false Questions

1. The active principles found in medicinal plants are from primary metabolic products.   **True**   **False**

2. The medicinal compounds of aloe concentrate in the sticky sap of leaves.   **True**   **False**

3. The doctrine of signatures was the principle source of reference for modern medicine.   **True**   **False**

4. Often it is the dose that determines the medicinal or toxic effects of mayapple.   **True**   **False**

5. Aloin is an alkaloid.   **True**   **False**

6. Treatment for dropsy used cyanogenic glycosides from foxglove.   **True**   **False**

7. Most glycosides have an active principle attached to a sugar group.   **True**   **False**

8. Hindu healers of India used snakeroot to treat lunacy.   **True**   **False**

9. A protozoan parasite causes malaria.   **True**   **False**

10. The use of ephedra is well known in cancer treatment.   **True**   **False**

CHAPTER 10
PLANT NARCOTICS AND POISONS

This chapter loosely applies the term "**plant narcotics**" to a variety of metabolic secondary plant products that have a profound effect on the human mental state, ranging from hallucinations to severe addictions. These substances consist of a wide variety of stimulating alkaloids that include the most dangerous natural narcotics. We also know them as psychoactive drugs; some are dangerously addictive and known to produce a variety of withdrawal symptoms.

These plant drugs bind to the nervous system and alter the natural interactions between nerve cells, which in turn causes changes in a person's mood, perceptions, behavior, and motor skills. One can administer such plant drugs in a variety of ways: by smoking, by applying directly on the skin, by inhaling (sniffing), or by injecting through a needle. The dosage may often determine the use of a drug for medicinal, psychoactive, or toxic purposes.

Depending on the dose, the administration methods, and the individual's tolerance to the drug, the effects on the human body and mind may vary. There are many ways in which psychoactive drugs can affect the biochemical processes in the human brain. Initially they produce a general sense of well-being by reducing tension and relieving pain. A drug overdose leads to confusion, convulsion, severe drowsiness, respiratory depression, and finally death. For example, cocaine causes psychological dependence by stimulating the brain's reward system; and opiates are physiologically addicting, which can produce a quite unpleasant withdrawal syndrome.

Historically, people associated plant drug use with religion and magic as well as medicine. Written documents attesting to the use of narcotics and other drugs are available from many cultures. Most countries still allow the practice of using plant drugs within religious, cultural, and medicinal contexts and consider such use non-recreational.

Classification of Plant Drugs

We can classify plant drugs by more than one scheme and provide one such classification below. When looking at the following classification, please be aware that some of these drugs may act in a combination of ways. For example, nicotine and opiates, classified as stimulants and depressants respectively, also may induce hallucinations in people.

Stimulants: Temporarily increase alertness, reduce hunger, and increase performance. *Examples:* Caffeine, Ephedrine, Nicotine.

Hallucinogens: Cause changes in thought, emotion; modify perception and consciousness; induce a dreamlike state and affect cognitive skills. *Examples:* Tropane alkaloids.

Depressants: Induce a sleeplike state, dull mental awareness; reduce physical performance; create loss of motor coordination and slurred speech. *Examples:* Opium alkaloids.

Plants Often Misused

Marijuana

Marijuana (ganga, pot, grass, weed) is the name for the crude drug made from the shredded plant parts of the hemp plant, ***Cannabis sativa***, Cannabinaceae. The plant originated in Central Asia and is widespread all over the world today. Cannabis plants are annual dioecious tall herbs that vary in size and form among different strains. The palmately compound leaves are attractive and very distinct. The unfertilized female flowers and tender leaves adjacent to them are covered by glandular trichomes with sticky resin, the source of the

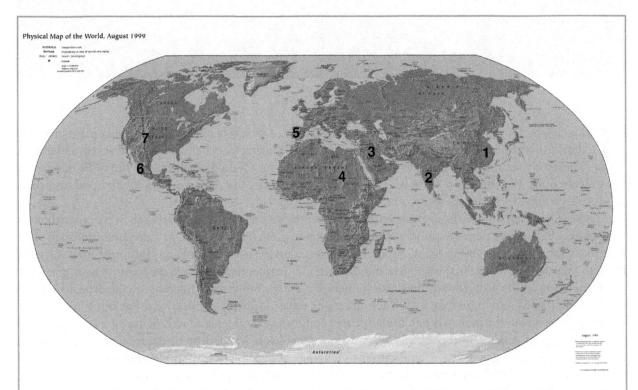

1. **China:** Used it to make first true paper in 105 CE; wore hemp clothes; used in medicine.

2. **India:** Hallucinogenic properties first exploited; attributed to Lord Shiva; prepared bhang.

3. **Arab world:** Associated with local legends; discovered hashish; largest user in early times by 1200 CE.

4. **Africa:** Introduced by Arabs by 1600 CE; used as a mild sedative during child birth; called it dagga; made a beverage; later smoked it in hookah.

5. **France:** Introduced through Napoleon's soldiers; others used it to enhance intellectual power; spread through the rest of Europe.

6. **Central America:** Introduced by Spaniards; used by African slaves who worked in sugar fields.

7. **North America:** Initially grown for fiber; drug use started after 1800s; used by jazz musicians.

Fig. 10.1 The path of marijuana throughout the world

psychoactive compounds known as tetrahydrocannabinol (THC). The types and parts of the plant where the drug came from, the soil conditions, the time of harvest, and many other factors determine the strength and quality of the drug.

Historically, the Chinese were the first to use hemp for medicinal purposes and for fiber. Archaeological evidences attest to the cultivation of hemp for the fibers since 8000 BCE. The Chinese called it "*ma*" and had high respect for the plant. They made the first true paper by 105 CE from the hemp fibers extracted from the plant. They could twist the durable fibers into ropes and spin it into clothing. The Chinese wore hemp clothing as a way of remembering and showing respect for the dead in the family. Because of this and the reputation for their silk, people referred to China as the "land of mulberry and hemp." They used hemp seeds as food and for oil used in paints and for lighting the lamps. They were aware of the dioecious nature of the plants and encouraged the cultivation of female plants that are rich in sticky resin.

The people of India first exploited the hallucinogenic properties of *Cannabis*. The Hindu Vedas describe one of the Hindu Gods, Lord Shiva, as the "Lord of Bhang." Indians were also aware of the dioecious nature of the plants and knew that unfertilized female flowers yielded a more potent resin. These plants were kept seedless and referred as **sinsemilla** (for *sin* without and *semilla* for seed) derived from American Spanish. To encourage the production of more female flowers and to delay their fertilization, they practiced special cultivation strategies. Indians recognized three grades of *Cannabis:* the least potent, dried leaves used in the preparation of a milk-based religious beverage called **bhang**; the potent female flowers and upper leaves, the **ganga** that they usually smoked; and the most potent, pure resin derived from glandular trichomes as charas. The Arabic term hashish refers to the charas of Northern India. Hashish may be solid or resinous which was used in hookas (water pipes) and smoked by Arab people.

Throughout the Middle East and Africa, marijuana use predominantly took place using hookas, a common practice in public markets and streets. Arab traders introduced the drug to Africa during the fifteenth and sixteenth centuries. In Africa, they used the plants medicinally to ease the pain of childbirth and to wean children. They also chewed dried plant parts and mixed them into beverages. Marijuana use spread throughout the western part of the world starting in the nineteenth century. In Europe, many people believed that Paris hashish clubs were the central intellectual meeting places. Members of this club were writers, poets, and artists who believed that hashish enhanced their creative abilities. Spaniards introduced marijuana to the New World mainly for hemp fiber production. By the turn of the twentieth century, marijuana found its way to the United States either from the Caribbean Islands or through Mexican immigrants who were already smoking it. Jazz musicians found a way to smoke marijuana in cigarettes, which quickly spread as a fad among fans and other public. Anti-marijuana campaigns began in the United States in the 1920s and spread through other countries without much success.

Heavy marijuana abuse results in psychological addiction leading to a habit that is hard to break. Heavy users experience reduced motor coordination, increased heart rate, anxiety, reduced sex drive, loss of memory, and distorted perception.

Opium

Opium is from the opium poppy, **Papaver somniferum**, Papaveraceae, a large annual herb with attractive, brightly colored flowers. It is native to southeastern Europe and western Asia, and many parts of the world grow it widely, including South America. India is the world's largest legal producer of opium. Sumerian tablets from 2500 BCE record its use as a "joy plant." Ancient medical records, including the works of Hippocrates, mention opium. The tender, green fruits called capsules exude a milky latex when slit. The brown, dried latex is the crude opium that contains about thirty different alkaloids, including morphine, codeine, and papaverine. It is also the source of semi-synthetic heroin that is at least six to ten times more addictive than morphine. Scientists generally recognize poppy seeds as safe for human consumption, and consumers use them in baked goods and for oil.

The Opium War had a major impact upon the history of China. Opium smoking became introduced to China around the 1600s along with tobacco smoking. Soon, tobacco use decreased and opium use increased.

Around the 1800s, Chinese silk, tea, and porcelain were in great demand among the British people. Instead of paying in silver, Britain traded opium for their desired goods from China by establishing opium plantations in nearby India. The Chinese opposed this opium trade by destroying opium in the Canton harbor. This started the Opium War in 1839, which led to China's defeat in 1842. As a result, Hong Kong became a British colony, and Britain established more opium ports for the trade. Meanwhile, opium addiction in China continued to rise. The second opium war started in 1865, and Britain again defeated China. Opium use among Chinese continued until 1913 when the opium trade finally stopped due to moral pressures from both countries. China strictly prohibited opium trade and use after the establishment of the People's Republic of China in 1949.

The pharmaceutical industry uses opium alkaloids as analgesics, sedatives, and pain killers. A German scientist, Frederick Serturner, isolated one of the alkaloids, morphine (named after Morpheus, the Greek God of dreams), in the early 1800s. As a purified alkaloid, people used morphine as an analgesic and a pain killer. An overdose of morphine could lead to death by suppressing the respiratory center in the brain. The Bayer company introduced heroin in 1898 as a semi-synthetic compound, which it believed was superior to both morphine and the least addictive codeine. Heroin addiction has increased dramatically over the past four decades and continues to be a serious problem. Opium and other drugs derived from opium are dangerously addictive and can have severe toxicological effects on the human body.

Fig. 10.2   A coca user depicted in an Incan artifact.

Coca

Coca, *Erythroxylum coca*, is the source of the powerful alkaloid cocaine. The word Coca is derived from an Aymaran word meaning "alimentation for workers and travelers". Coca plants are native to the humid eastern slopes of the Andes Mountains in South America. Archaeological evidence supports the use of leaves among inhabitants of Ecuador, Peru, Bolivia, Colombia, Argentina and Chile for at least four thousand years. Within the Inca Empire, coca leaves were considered sacred and every ceremonial offering to the Andean Gods included burning the coca leaves. They used the leaves in place of currency in exchange for other food items. The natives chewed the leaves for their mild stimulatory effects. A normal practice of chewing the leaves involved mixing the slowly dried leaves with a catalyst which helped extract and absorb the alkaloids more efficiently. The choice of catalyst varied among different ethnic groups but always included a plant ash of alkaline nature. The chewing practice itself was a slow process that involved conversation and friendship. The ball of quid from the chewed leaves was removed by hand as a form of respect as opposed to spitting it out. The leaves are best when used as above and lose their potency in storage.

Cocaine (also called Blow, Coke, Crack, Snow) is an euphoric stimulant and appetite suppressor. When the Spanish conquerors allowed the enslaved native Indians to chew coca leaves, the natives' endurance and productivity increased in the field. Cocaine's popularity has increased since 1860, when Albert Niemann first isolated it in a laboratory. People who used it initially for the treatment of colds, asthma, and hay fever became strongly addicted. A cocaine-based beverage coca-wine, prepared by an enterprising Italian, Angelo Mariani, soon became a rage in Europe and gained popularity among celebrities by the late nineteenth century. Another beverage that included cocaine as one of the main ingredients was Coca-Cola, prepared in 1886 by John Styth Pamberton, an Atlanta pharmacist. After the company realized the addictive properties of

Fig. 10.3a Cannabis
© Dimon044/Shutterstock.com

Fig. 10.3b Peyote
© Todd Pierson/Shutterstock.com

Fig. 10.3c Mandrake
Courtesy Rani Vajravelu

Fig. 10.3d Belladonna
© Sue Robinson/Shutterstock.com

cocaine in 1903, they removed it from the recipe. As a result, today the beverage contains coca leaves with the cocaine removed.

There are several forms of cocaine. One is a hydrochloride salt known as **coke** that dissolves in water. One can inject it into a vein or snort it through the nose. **Crack** is the smokable form of cocaine produced by treating hydrochloride with boiling water and baking soda. This form of cocaine comes in a rock crystal, hence it is also known as "**rocks**." Free base is mostly pure cocaine that can reach the brain within fifteen seconds. Depression, anxiety, and hallucinations usually follow the "highs" the user may experience. Regardless of how one uses cocaine, a user can experience several health problems, including heart attack, respiratory failure, strokes, and digestive problems that can become fatal. Cocaine was the drug of use among celebrities in the United States until the 1980s when cocaine-related deaths raised concern among the public.

Tobacco

Tobacco yields an alkaloid, nicotine, which ranks next to alcohol and caffeine among people who use them around the world. Two common species that contain high amounts of nicotine are *Nicotiana tabacum* and *N. rustica* known in South America and the south Pacific. Another species, *N. suaveolens*, exists in Australia. The perennial, herbaceous plants grow as annuals and contain alkaloids in all parts of the plant, although mostly concentrated in their leaves. The roots produce nicotine, from where it is translocated to the leaves for storage. Native American Indians have used tobacco leaves since prehistoric times. They wilted, dried, and rolled the leaves to make crude forms of cigars or pipes for smoking. Native tribes have considered tobacco as a sacred plant and thought that the smoke carried messages to God. Amazonian tribe leaders have used tobacco in divination rites that are still in practice today. People have used tobacco to ease the pain of childbirth and to suppress hunger. During Columbus's first voyage to the New World in the 1400s, native tribes offered him the "peace pipe" as a symbol of hospitality, and he took some back to Queen Isabella. Tobacco use spread through France after the French Ambassador, Jean Nicot, popularized it in the late sixteenth century. Soon its use spread throughout the world, encouraging cultivation in British colonies to ensure a national supply. George Washington and Thomas Jefferson, former presidents of the United States, cultivated tobacco and also used it in place of cash for goods. The Jefferson Memorial in Washington, D.C. has a statue of Jefferson standing on a pedestal of tobacco and corn.

Growers can raise tobacco plants from seeds. Once they are ready to flower, they top the plants to encourage leaf expansion. The cut leaves go through a curing process by wilting, slow-drying, and fermentation processes. Consumers can roll such cured leaves into cigarettes or cigars for smoking or cut them into pieces or grind them up to chew or use as a snuff. Nicotine is a stimulant of the central nervous system, and we now know it is one of the most physiologically addicting drugs. It raises blood pressure, slows the heart rate, and has links to various forms of cancer. Despite the health concerns, however, the tobacco plant still remains the most important cash earning, non-food crop in the world.

TABLE 10.1 Most popular Plant Narcotics

Psychoactive Nature	Common Name	Binomial	Psychoactive Compounds	Family
Depressant	Opium	*Papaver somniferum*	Opium alkaloids	Papaveraceae
Hallucinogen	Marijuana	*Cannabis sativa*	Cannabinoids	Cannabinaceae
Hallucinogen	Belladonna	*Atropa belladonna*	Tropane alkaloids	Solanaceae
Hallucinogen	Henbane	*Hyoscyamus niger*	Tropane alkaloids	Solanaceae
Hallucinogen	Mandrake	*Mandragora officinarum*	Tropane alkaloids	Solanaceae
Hallucinogen	Jimson weed	*Datura stramonium*	Tropane alkaloids	Solanaceae
Stimulant	Coca	*Erythroxylum coca*	Cocaine	Erythroxylaceae
Stimulant	Tobacco	*Nicotiana tabacum*	Nicotine	Solanaceae

Deadly Nightshade and Related Plants

Tropane alkaloids (also known as belladonna alkaloids) are a class of secondary metabolites naturally found in many members of the Nightshade family, Solanaceae. Belladonna, Henbane, and Mandrake contain relatively high concentrations; all three species have a long history of use as magic plants connected with witchcraft and superstition. Hallucinations derived from these tropane alkaloids include visions of wolves.

Belladonna, ***Atropa belladonna***, yields atropine that one can absorb directly through the skin. Witches in the Middle Ages rubbed the ointment mixed in oil on their bodies. The hallucinogenic sensations produce a psychological sensation of flying. Atropine induces a deep sleep and causes the user to lose all sense of reality. Atropine and its synthetic derivatives could be extremely toxic.

Henbane, ***Hyoscyamus niger***, contains scopolamine, hyoscyamine, and atropine, and is the least toxic due to relatively low amounts of atropine. However, when ingested in large quantities it can cause hallucinations, anxiety, convulsions, impaired vision, and death. The Egytian scroll, Ebers Papyrus written in 1500 BCE mentions the use of Henbane in the preparation of magic drinks.

Mandrake root, ***Mandragora officinarum***, has a morphological resemblance to the male and female human figure. Hence, ancient people believed it had supernatural powers over the human body and mind. The large brown root resembles a parsnip, either entire or branched out. The roots contain mandragorine, which researchers believe is very similar to atropine or hyoscyamine.

All of the above three played a great role in Europe during the Middle Ages. The ancient use of these alkaloids in folk medicine and witchcraft led to their medicinal use.

Jimsonweed, ***Datura stramonium***, is native to the United States, although other species are common in Europe and Asia. People have used this plant as an intoxicant for several centuries. The plants are rich in scopolamine, which is the most hallucinogenic of all tropane alkaloids. They are very popular among young teenagers who are unfamiliar with the dangers of these alkaloids in high doses.

Angel's Trumpet, ***Brugmansia suaveolens***, is a perennial, short tree with large and attractive trumpet-shaped, pendulous flowers. It is native to southeast Brazil. All parts of this plant are toxic in large doses due to the presence of scopolamine and atropine. Indigenous people of Peru and Brazil have used the plant extracts as recreational drugs to induce hallucinations. The common name, Angel's Trumpet is mistakenly used for *Datura* as well due to the presence of large trumpet shaped flowers.

Lesser Known Plant Drugs

Peyote

Peyote (or mescal buttons), ***Lophophora williamsii***, Cactaceae, is one of the new world hallucinogens. The Indians of Mexico have used it for at least two thousand years before Europeans arrived. Its use has spread to North American Indian tribes in the last hundred years. The woolly, green cactus grows close to the ground and contains the most active hallucinogen mescaline among thirty different alkaloids. Slicing and drying this plant does not cause it to lose its hallucinogenic properties. When needed, these slices (that look like buttons) can be either soaked in the mouth and then the saliva with peyote chemicals swallowed. One can also soften it in water to release the alkaloids and then drink it. Peyote is a narcotic that induces hallucinations that can last up to twelve hours. During this period it produces kaleidoscopic illusions of vivid colors and distorted perceptions of time. Some people report hearing the voices of their ancestors. There are other reports that describe anxiety and visions of wild animals, including snakes. In some people it induces nausea, tremors, chills, and vomiting. Despite the condemnation of Spanish conquerors in the early times and local governments now, the practice of using peyote in religious ceremonies has continued over the years among Mexicans and is continuing in Native American Churches today.

Kava

Kava refers to a shrubby plant, *Piper methysticum* (*methysticum* in Greek means intoxicating), and also a pungent beverage made from its roots. This plant is in the family Piperaceae, the family of black pepper. Kava is an ancient crop of the Pacific Islands, and its use extends throughout Polynesian cultures all the way to Micronesia and Australia. Most parts of the Pacific islands used kava as a social, ceremonial, and sacred beverage. Kavalactones, the active ingredients, mostly accumulate in their stumpy roots. The beverage is prepared by chewing the root fragments by mouth a few at a time, collecting them in a bowl, and mixing them with water. The mixture is usually strained using coconut fiber net. They consume the beverage as quickly as possible in a coconut shell. Modern preparations involve grinding kava roots and mixing them in water. Traditional chewing practices yield more potent kava. The taste and aroma vary between fresh and dry roots and depending upon the plant variety. The beverage is a tranquilizer and a muscle relaxant and believed to be non-habit forming. Excessive consumption may lead to significant skin problems and loss of appetite. Kava is sold in the form of pills, extracts, and powder in modern health food stores to relieve stress, insomnia, and anxiety.

Betel and Areca Nut

Another plant from the pepper family is *Piper betel*, a perennial climber with glossy, crisp-textured leaves, also known as *paan* leaves. The plant is native to India and Sri Lanka where the people have chewed the leaves as a tradition after meals as a relief for a range of ailments, from dry mouth to digestive problems. To start the chewing practice, they wrap a single leaf with pieces of Areca palm nuts and masticate them slowly until the saliva turns red and the lips and tongue are stained. They add lime (calcium hydroxide) to this wrap to enhance the release of alkaloid from the betel nut. They spit out the saliva periodically and finally spit out the masticated quid. In Hindu religious ceremonies, betel leaves and *Areca* nuts are offered to the god, to the priest, and to the guests. The leaf contains betel oil, which is supposedly a stimulant, a mild aphrodisiac, and a breath freshener. People include the Areca nut for the energy boost derived from its natural chemicals. They consider the chewing of paan a social practice as well as a gesture of friendship and companionship.

Ololiuqui

Ololiuqui is the morning glory, both *Rivea corymbosa* and *Ipomoea violacea* and other related species of the genera from Convolvulaceae. Ololiuqui is the Aztec name for the seeds that local tribes have used in religious ceremonies and as an ingredient in magical ointments since prehispanic times. The plants are vines with thin cordate leaves and long, funnel-shaped flowers. The seeds contain various amides of lysergic acid. The traditional method of preparation is to soak the crushed seeds in water for several hours and then consuming both the seeds and water. The tribal leader would consume this magical beverage to induce hallucination. Under intoxication, the leader would receive messages from the supernatural and uncover the cause of a person's illness. The natives believed in the medicinal properties of the seeds to relieve pain and tumors among others. The natural plant compounds are less potent compared to synthetic LSD.

Khat

Khat is from *Catha edulis* of Celastraceae, a species native to Ethiopia. People chew the fresh leaves with lime (calcium hydroxide) until masticated and the juice releases. They then swallow the remaining cellulose mass. It is a stimulant and may become addictive. The use of khat is common and accepted among African Muslims.

Caapi

Caapi, also known as ayahuasca, yaje, or cipo, from a few related species of ***Banisteriopsis*** included in Malpighiaceae. The term Ayahuasca comes from the Quechua language meaning "rope of the spirit". It is an important plant for the South American Amazonian tribes. They use infusions of mashed bark or stem or just chew it to release hallucinogenic compounds. Different plants including *Psychotria* spp. are also routinely added to this preparation to enhance the effect. The reactions vary from being pleasant to having terrifying experiences of seeing wild animals, such as jaguars or snakes. Use of caapi is common in communal rituals.

Poisonous Plants

Castor beans, the seeds of ***Ricinus communis*** from the Euphorbiaceae, are poisonous to people, animals, and insects. Poisoning by the castor bean seed is due to a toxic protein, ricin. If people consume the seeds, symptoms may delay up to several hours or a few days. The initial symptoms of severe dehydration, nausea, diarrhea, vomiting, and low blood pressure are followed by internal hemorrhaging, kidney failure, and damage to the liver. It is the most deadly natural poison known to humankind. As small an amount as a milligram could kill an adult.

Milkweeds, *Asclepias* spp. and **Oleander,** *Nerium oleander* from Apocynaceae, contain cardiac glycosides that act very much like digitoxin that comes from the foxglove plant. Many common household aroids, such as dumb cane, ***Dieffenbachia*** spp., philodendrons, contain calcium oxalate crystals in the form of needles that result in painful burning and swelling and may cause temporary paralysis of the throat muscles. Other equally dangerous aroids are the jack-in-the-pulpit, ***Arisaema triphyllum***, and skunk cabbage, ***Symplocarpus foetidus.***

Curare refers to different types of concoctions made from poisonous plant extracts and other additives that may include snake venom. Natives in tropical South America use curare as arrow poison. The most commonly used plants are ***Strychnos toxifera*** of Loganiaceae and ***Chondrodendron tomentosum*** of the Menispermaceae. Both plants are commonly referred as curare plants, which contain many alkaloids in their bark juices, including curarine and tubocurarine that affect neuromuscular transmission. Curare causes the paralysis of skeletal muscles and may lead to death if immediate treatment is unavailable. Alkaloids from curare plants are used

TABLE 10.2 Poisonous Plants

Common Name	Binomial	Family	Poisonous Compounds
Castor bean	*Ricinus communis*	Euphorbiaceae	Ricin
Curare	*Strychnos toxifera*	Loganiaceae	Curarine; Tubocurarine
Curare	*Chondrodendron tomentosusm*	Menispermaceae	Curarine; Tubocurarine
Dumbcane	*Dieffenbeckia spp.*	Araceae	Calcium Oxalate Crystals
Jimson weed	*Datura stramonium*	Solanaceae	Tropane alkaloids
Mescal bean	*Sophora secundiflora*	Fabaceae	Cytosine
Milkweed	*Asclepias spp.*	Apocynaceae	Cardiac glycosides
Nux vomica	*Strychnos nux-vomica*	Loganiaceae	Strychnine
Oleander	*Nerium oleander*	Apocynaceae	Cardiac glycosides
Poison hemlock	*Conium maculatum*	Apiaceae	Coniine

in medicine as a muscle relaxant during surgery as well as in the treatment of acute arthritis, spastic cerebral palsy, myasthesia gravis, polio, and tetanus.

Strychnos nux-vomica, a plant native to southeast Asia, is one of the deadly poisons known to humans. The alkaloid strychnine found in its seeds causes convulsions of all voluntary muscles and a drastic change in blood pressure that leads to death. People routinely employ it as rat poison.

Poison hemlock (also known as poison parsley) is in the same family as carrots and celery. The most familiar plant is *Conium maculatum* that is native to Europe and the Mediterranean region and naturalized in North America, Australia, and most parts of Asia. This herbaceous plant has finely dissected leaves and white flowers clustered in umbels. The poisonous alkaloid is coniine, which is a neurotoxin. When ingested, even in a small quantity, it disrupts the central nervous system and causes respiratory muscle paralysis to both people and livestock. A similar plant from the same family is water hemlock that exudes a yellow, highly toxic sap from the cut roots that will cause violent convulsions leading to death.

Mescal bean is a member of Fabaceae, one of the major sources of alkaloids. The bright red seeds of *Sophora secundiflora* (mescal bean) are a source of hallucinogen used by North American Indian tribes. In slight overdoses, the highly toxic alkaloid cytosine could cause paralysis of respiratory muscles and result in death. Another plant commonly found in the southeastern United States is the jequirity bean, *Abrus precatorius*, which contains the toxic abrin in the glossy, attractive, red seeds that can become highly toxic when chewed.

Review Questions

Give brief answers in the given space.

1. Classify the plant narcotics based on their effects on the human body.

2. List five plant families that are rich in psychoactive plant compounds.

3. Narrate the early uses of *Cannabis* in China.

4. What is Ayahuasca? Briefly state its sources and uses among the natives of South America.

5. Describe the events that led to the Opium War.

6. Discuss the ancient use of Coca among the native people of the Andes mountains.

7. Explain why peyote is also known as mescal button.

8. List the plants that are rich in tropane alkaloids.

9. What is curare? List two plants used in curare preparation.

10. Explain the toxic nature of castor beans.

Multiple Choice Questions

For each question, choose one best answer.

1. The psychoactive compounds of marijuana mostly occur in the:
    A. capsules
    B. latex
    C. seeds
    D. glandular trichomes
    E. mature leaves

2. The largest legal producer of opium is:
    A. The People's Republic of China
    B. India
    C. The United States
    D. Australia
    E. None of the above

3. Frederick Serturner isolated which of the following?
    A. Cocaine
    B. Codeine
    C. Morphine
    D. Atropine
    E. Nicotine

4. Which of the following would fit in the category of stimulant drug?
    A. LSD
    B. Opiates
    C. Marijuana
    D. Cocaine
    E. Mescaline

5. Which of the following belongs to Convolvulaceae?
    A. Peyote
    B. Kava
    C. Ololiuqui
    D. Kat
    E. Caapi

Match the following chemicals from Column 1 to the appropriate plant source in Column 2. You may use a choice once, more than once, or not at all.

Column 1

6. Abrin

7. Cardiac glycosides

8. Coniine

Column 2

A. Milk weed

B. Poison hemlock

C. Belladonna

D. Jequirity bean

E. Mescal bean

9. Which of the following genera exemplify the doctrine of signatures?

   A. *Hyoscyamus*
   B. *Mandragora*
   C. *Erythroxylum*
   D. *Lophophora*
   E. *Nicotiana*

10. Which plant drug was traditionally prepared by chewing the root fragments to make a beverage?

   A. Caapi
   B. Kat
   C. Kava
   D. Tobacco
   E. Jimsonweed

True/false Questions

1. A legal drug obtained from opium is codeine.   **True**   **False**

2. *Cannabis* is native to the New World.   **True**   **False**

3. The most recent discovery of opium use was in the Arab world.   **True**   **False**

4. Crude opium is derived from the dried latex.   **True**   **False**

5. Scopolamine is a tropane alkaloid.   **True**   **False**

6. Native American churches accept peyote use.   **True**   **False**

7. India is the "Land of mulberry and hemp."   **True**   **False**

8. Crude opium traditionally filled "hookas."   **True**   **False**

9. Roots originally produced the alkaloids in tobacco.   **True**   **False**

10. Ricin is an alkaloid.   **True**   **False**

CHAPTER 11
PLANT FIBERS AND WOOD PRODUCTS

A s you begin to read this chapter, take a minute and look around. Start from the paper this book is printed on, move on to your clothes, the chair, the desk, the cushion or sofa you are using, the floor or the mat on it, the panels on the wall or the draperies in the room, and finally the door or the fence outside if you are sitting near a window. What is common to all of these items that surround you? You probably guessed it already. If not, read on. Most of the above items either use plant fiber or a wood product even if you cannot readily make out. This is the twenty-first century; we live in concrete buildings and communicate through technology. Despite this, look how many natural plant products make up the place you live in. It is a known fact that in prehistoric days plants provided the shelter, clothing, and fuel for humans. Compared to a few hundred years ago, the use of plant resources has only increased, as you will find out in this chapter. Many items we use every day that make our lives livable and comfortable either directly or indirectly use plant fiber and wood.

Anatomy of Fiber and Wood

Fiber and **wood** are certain cell and tissue types of plants. Chapter 2 discussed basic tissue types. Recall the sclerenchyma fibers and components of xylem tissue. The cells of sclerenchyma fibers and the components of xylem tissue, such as vessels, tracheids, and fibers, are basically composed of a cellulosic cell wall that lignin further thicken to form a secondary wall. The living protoplasm is lost in this process, which turns them into dead cells that provide mechanical support to the plant parts. Xylem tissue is involved in conducting water and dissolved minerals throughout the plant. As the plant undergoes secondary growth, the accumulations of secondary xylem form the wood in a plant.

The chapter will focus on the distribution, extraction, uses, and some sources of fiber and wood from plants. The first part focuses on the fiber; the second part discusses wood and its commercial products.

Fiber

Fiber, the word we use almost everyday, may mean different things based on the situation. In human diet it means the indigestible cellulose that adds to the bulk of the food when eaten in the form of whole grain or salad. In botany it refers to the elongated, tapering, thick-walled, dead cells called sclerenchyma fiber. In the textile industry it refers to collections of sclerenchyma fiber cells, or the entire vascular bundles used for fabrics, ropes, and mats.

Fig. 11.1a Fibers
Courtesy Rani Vajravelu

The composition of the cell wall in most fibers is chiefly cellulose, although lignin, as well as tannins, gums, pectin, and other polysaccharides, may also be present. The most valuable fibers are those that are nearly pure cellulose and white in color. Fibers that are high in lignin are generally of poorer quality, browner in color, and of lower mechanical strength. Cellulose is an extremely strong material. It determines the tensile strength of a fiber—the extent to which one can stretch it before it breaks. Plant fibers composed of cellulose do not denature by high temperatures.

As we have seen in Chapter 2, sclerenchyma cells are in the ground tissue or along with xylem and phloem tissue. In vascular bundles of dicot stems, mature phloem turns to phloem fiber and occurs in patches over the living phloem cells (see Figure 11.1b). Cotton is an exception, where the fibers exist as outgrowths on the seed epidermis.

Based on the plant part from which they derive, they become categorized as surface fibers, soft fibers, and hard fibers. One must remove the fiber cells and clean away the rest of the plant tissue to make them usable in textile or other commercial applications. Manufacturers use standard methods for all plants except for surface fibers like cotton. We discuss the most common fiber extraction methods below.

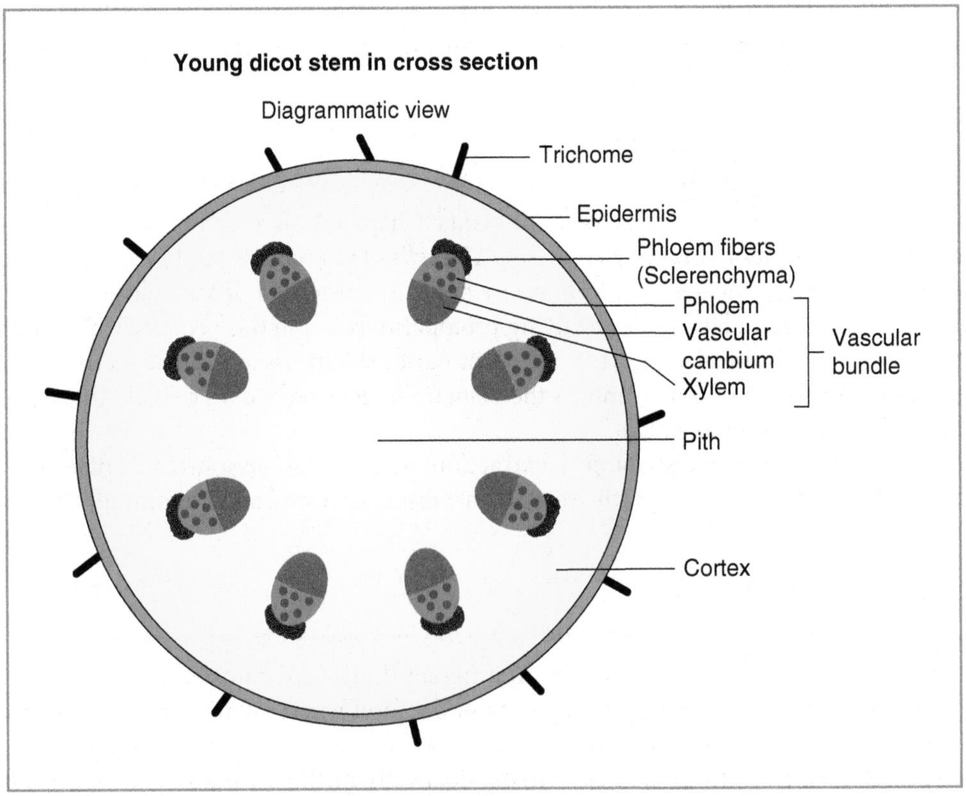

Young dicot stem in cross section

Diagrammatic view

- Trichome
- Epidermis
- Phloem fibers (Sclerenchyma)
- Phloem
- Vascular cambium
- Xylem
- Vascular bundle
- Pith
- Cortex

Fig. 11.1b Location of various tissues in the stem
Courtesy Rani Vajravelu

Extraction methods

Ginning is a process employed for removing seed fibers. Derived from the word for engine, gin means machine. Using a gin, workers can quickly and easily separate the seed fibers, which otherwise would be time consuming and cumbersome. They further clean, tease, card, weave, and bleach the extracted fibers before they make it into stores.

Retting is the initial process after harvesting the fiber plants. This process involves water and microbial action on the soft parenchyma tissue and other material such as gum and pectin that hold fiber cells. The workers soak the plant stems in water for a period of time and constantly check to make certain that only the unwanted tissue and not the fibers have decayed. They use retting for all soft fibers, such as flax, hemp, jute, and ramie.

Scutching is a process that follows retting. It involves beating and scraping the fibers to remove the remaining unwanted tissues, such as xylem, which adhere to the fibers. After removing the soaked plants from the water, followed by washing and drying, the workers crush the plants under fluted rollers. This breaks the dry woody matter sticking to the fibers. **Hackling** is aligning the   separated fibers in vertical pins to detangle them. **Decorticating** is a method mainly used for leaf fibers. The workers crush the fleshy leaf material and scrape the non-fibrous materials.

Some of the most commonly known fiber sources are discussed next.

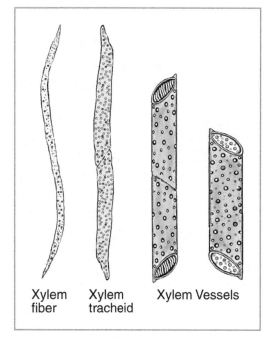

Xylem   Xylem   Xylem Vessels
fiber   tracheid

Fig. 11.1c Xylem cell types
Courtesy Rani Vajravelu

Surface Fibers

Coconut

Coconut, *Cocos nucifera,* Arecaceae, yields **coir** fibers from the mesocarp of the fruit. The bulk of an entire coconut fruit consists of rough, stiff, brown, relatively short fibers. We obtain coir as a byproduct of copra and the coconut oil industry. The manufacturer strips off the mature coconut mesocarp along with the exocarp and retts them for nearly ten months in water. Retting softens the pulpy remains that the workers then beat and wash in order to remove the unwanted tissues and to retain the fibrous part. The clean fibers are still brown due to the high lignin content that are dried, spun into yarns. Coconut fibers are used in ropes, matting, shell fish netting, plant baskets, and so forth. One obtains the best quality coir from immature fruits; coir obtained from mature coconuts has a rougher texture. Such mature coir fibers are decorticated after retting, which separates two types of fibers. One type is the shorter fibers used in mattress stuffing; the other longer type is bristle fiber that has application in door mats, brushes, and brooms.

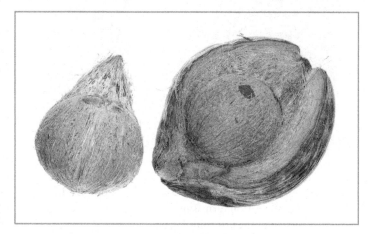

Fig. 11.2a Shelled coconut and the mesocarp fibers
© sunsetman/Shutterstock.com

Cotton

Several species of the genus *Gossypium* from Malvaceae produce the cotton of commercial value. This genus includes about thirty species ranging from herbaceous to small tree forms. There is a considerable amount of information and confusion about the wild ancestors of modern cotton plants. Botanists generally agree upon four distinct cultivated species: diploid *G. arboreum* and *G. herbaceum* that are from the Old World and known for their short lint fibers called **staples**; and tetraploid *G.hirsutum* and *G. barbadense* that are from the New World. Due to the quality and length of their fibers, the cotton industry prefers these and grows them extensively.

G. arboreum, also known as **tree cotton**, is a native of India and Pakistan. The adjacent tropics also domesticated it. Historical documents indicate its cultivation four thousand years ago by Indus Valley civilizations.

G. herbaceum, also known as **Levant cotton**, is native to Africa and Arabia. Originally Ethiopians cultivated it; then later it spread to Iran and the rest of the Near East. By the mid 1400s both species had become introduced to the rest of Europe.

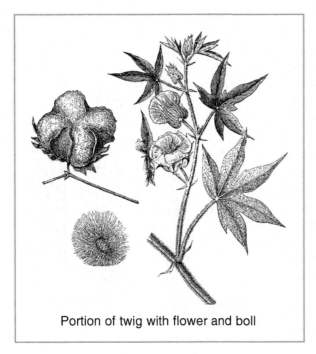

Portion of twig with flower and boll

Fig. 11.2b Cotton plant
© Hein Nouwens/Shutterstock.com

G. hirsutum, also known as **upland cotton** or West Indian cotton, has appeared in archaeological excavations in Mexico dated 3400 BCE. It is the number one cotton source in the world, accounting for more than 90% of the world's cotton.

G. barbadense has appeared along the coastal regions of Peru since 2500 BCE. Western South American cultures practiced cotton weaving as an important part of their lives. This species appeared in the West Indies before the arrival of European colonists in the New World. Initially growers planted it along the coastal regions of the southern United States and referred to it as **Sea Island cotton**. Some plants of this species were introduced to Egypt, hence the name **Egyptian cotton**. The cultivars of this species yield relatively long, soft lint fibers that provide the finest quality threads. American **pima cotton** is another cultivar of the same species that growers planted in Pima County, Arizona.

Cotton plants have spherical or ovoid, leathery capsules called **bolls**. These are about three inches long and split at maturity into several valves, exposing the seed hairs called **staple**. It takes about two months for an open flower to produce a fully ripe open boll. There are many seeds, which are oval, solid, and up to a half inch long in upland cotton. They are usually smaller, grayish black, and hidden under one or two coats of hairy outgrowths, staple (trichomes), which arise from the seed epidermis.

Fig. 11.3 A spinning wheel of the sixteenth-century used to wind and twist the cotton yarn. This European invention was an improved version of the first spinning wheel invented in India around 750 CE.

National Archives and Records Administration.

There are two types of staple: short, thick-walled hairs called **linter** that are hard to pull off the seed, and long, soft, pliable hairs called **lint** that easily detach from the seed and that one can twist or spin into threads. We call such fibers **textile fibers**. Linters are used in papermaking, and lint has applications in the textile industry. Lint fibers contain cellulose in the secondary wall that makes the fibers versatile and easy to process. Cotton fibers accept dyeing easily compared with other plant fibers, and they also withstand high temperatures.

Human selections have resulted in annual plants with short stature and uniformity in plant size, flowering, and fruiting times. Artificial sprays that retain only the bolls in the plant for machine harvesting initiate defoliation. Workers must clean out the harvested bolls to remove the seeds and then remove the staple from the seeds, which they call ginning. Eli Whitney's cotton ginning machine invented in 1794 accounted for a steep increase in the export of cotton within ten years. Seeds with shorter staple use **saw gins**; seeds that have long staple use **roller gins**. Workers collect the separated lint staple and compress it into **bales** (tight packages for market). Each bale weighs about five hundred pounds.

In textile mills, a machine opens the cotton bales. Then a mixing machine pulls the fibers and combs them in parallel lines to form a thin web called carding. The web of fine fibers is then made into untwisted loose strands of about an inch in diameter called **sliver** and sent to the combing machine. The combing process removes the shorter hairs and impurities. Slivers are then drawn into thinner strands to improve the strength and wound into bobbins and sent to the spinning process.

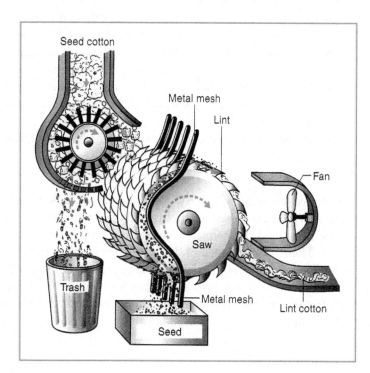

Fig. 11.4 A diagrammatic representation of the cotton gin operation. The rolling saw removes the lint from the seeds.
From *Plants and Society*, *5e* by Estele Levetin & Karen McMahon, 2008. Reproduced by permission of The McGraw-Hill Companies.

Since cotton is a plant fiber, it has natural pectin and waxes that manufacturers must remove to improve its texture and color. A process called **mercerization** accomplishes this by boiling the stretched threads in caustic soda for several hours and bleaching them in hydrogen peroxide. Invented by John Mercer in 1844, this process improved the cotton's luster, durability, and dye acceptance. The addition of a gel or starch improves the strength of the threads by making them stiff and suitable for the weaving process. Once the fabrics are made they become sanforized to reduce the shrinkage. This **sanforization** takes place by soaking in ammonia, which swells the fibers and prevents shrinkage after washing. Invented in 1970, this method has helped customers not to have to buy cotton clothes one size bigger in anticipation of shrinking after the initial wash.

Permanent press cotton fabrics are made by chemical processes to keep the cellulose fibers in shape. This prevents wrinkles after washing.

People in Egypt, Peru, India, and Pakistan have cultivated **natural cotton** that produces fibers in a variety of natural colors since 2700 BCE. These colors come in brown, tan, cream, green, or mauve, Due to the industrial revolution, which opted for white fibers, people ignored these naturals for some time. The colored cotton fiber plants are seeing a period of revival recently. No bleaching, dyeing, or chemical processing is necessary when one uses naturally colored fibers. Levi-Strauss has used colored G. hirsutum fibers in its line of jeans called "naturals."

Genetically Modified cotton was first developed in 2002 with the hope of eliminating pesticides on cotton plants. Initially it was unsuccessful due to poor yield. Growers reintroduced it in 2004. India is the largest grower of GM cotton.

Organic cotton is cotton grown without insecticides and harmful chemicals. The organic cotton is independent of Genetically Modified cotton. Organic cotton requires biological pest control and crop rotation. Due to the cost involved, organic cotton is expensive. Madhya Pradesh in India currently grows organic cotton.

Cotton facts

- Cotton is a major agricultural commodity that is not food-related.
- A cotton garment was worn only by the High Priest in ancient Egypt.
- The word cotton derives from the Arab word "Qutun."
- The invention of the ginning machine was the basis of the Industrial Revolution.
- Each cotton bale shipped to textile mills holds enough cotton to make 325 pairs of jeans.
- "Khaki" means dust color in India. Manufacturers now use it for jeans and clothes with a dull yellow-brown color.
- Cotton seed oil and seed residue are other important products from cotton.

Kapok

Kapok trees are from Malvaceae. We know **java kapok** or ceiba by the binomial *Ceiba pentandra*. It is native to Mexico and central and northern South America. You can easily identify the trees by the short, sharp spikes on the trunk and the palmately compound leaves. The tree is associated with the legends of Mayan civilization. The fruits are capsules that split open upon maturity exposing the soft, silky, fluffy, flammable, dull white fibers that surround each seed. One cannot spin the soft fibers into a thread because they are slippery. Instead consumers use them as stuffing for pillows, mattresses, stuffed animals, insulation, and upholstery. Java, Indonesia grows the commercial trees extensively. **Silk-cotton** or Indian kapok, *Bombax malabaricum*, yields a similar surface fiber used in similar ways mostly for local uses.

Bast Fibers

Bast fibers or soft fibers derive from thick-walled phloem cells (sclerenchyma fibers) that may range from six to fifteen feet long. Soft fibers derive from some dicot plants like flax, hemp, and jute. Retting separates these long fibers from the plants, followed by scutching. Even though manufacturers can bleach and dye most bast fibers, they cannot mercerize or sanforize them like cotton.

Fig. 11.5 Kapok tree
© Dr. Moreley Read/Shutterstock.com

Jute

Jute is the second most produced fiber next to cotton and is the world's most used bast fiber. The cost of production and the ease of fiber extraction have made this fiber very popular. Jute can come from more than one species; the most common one is *Corchorus olitorius*, Tiliaceae, which is native to the Mediterranean region. The plants are herbaceous annuals reaching an average height of fifteen feet. Jute fibers contain cellulose and lignin, which results in a dull white fiber. People from ancient times have used this plant for fiber and mainly for sack making, which still continues today. People throughout the Middle East and the Far

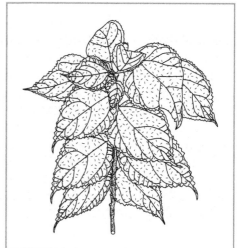

Fig. 11.6a Flax
© Hein Nouwens/Shutterstock.com

Fig. 11.6b Jute
Courtesy Rani Vajravelu

Fig. 11.6c Ramie
Courtesy Rani Vajravelu

East cultivate jute, with India being the primary producer. Jute fibers do not stretch and disintegrate in water. Consumers their fibers in canvas, carpet backing, twine, sacks, conveyer belts, and wall coverings. Particleboard, paper, and rayon use its by-products.

Linen

Linen or **flax fibers** are from *Linum usitissimum,* Linaceae, which is native to Europe and eastern Asia. Growers base jute cultivations upon domesticated species around the world. They grow modern cultivars either for fibers or for oil seeds. Flax is the oldest textile fiber known to humans since 8000 BCE. The Egyptians wrapped their mummies in linen fibers. The ancient Greeks and Romans used linen for its luster and beauty and spread it through Europe. The word *lingerie* derives from the Latin word *linen.*

The plants yield shorter fibers than jute but stronger fibers than cotton. Flax is the strongest of all vegetable fibers. It is a highly absorbent, good conductor of heat and remains cool to the touch. Due to its poor elasticity, linen clothes wrinkle easily. Because of its strength and durability, it is used in making button holes, mail bags and hoses. The use of linen in textiles has increased over the past thirty years. Linen fibers are used in bed sheets, tablecloths, window treatments, clothing, luggage, sewing thread, and so forth. The extraction of linen fibers requires the use of manual labor, which adds to the price of the fiber. It is expensive fiber. Linseed oil is extracted from the seeds.

Hemp

Hemp is *Cannabis sativa,* a well-known medicinal and psychoactive plant from Cannabinaceae. Hemp fibers derive from industrial hemp (grown for non-drug use). People have used hemp fiber since prehistoric times. Native to Central Asia, the Chinese used hemp extensively 4000 BCE. The plants are dioecious, annual herbs. Growers cultivate two separate subspecies: one for fiber and one for recreational and medicinal drugs. Hemp fibers resemble flax fibers except they are much longer (up to fifteen feet) and stiffer. After processing the fibers are silky, soft, and creamy white. Fibers are used in cordage, sailcloth, rope, canvas, biodegradable plastics, and health food. They often mix hemp with cotton, silk, and linen to make a blended fabric and used it in papermaking until wood pulp took over. When Thomas Jefferson wrote the Declaration of Independence, hemp fiber paper was used. Lately, France has become involved in hemp production. One of its products is a construction material called Hempcrete that uses hemp fibers and cement. People also eat the seeds and use them for a drying oil in paints.

Ramie

Ramie, *Boehmeria nivea*, Urticaceae, is also called China grass. It is native to the tropical parts of eastern Asia. The plants are herbaceous perennials reaching to a height of over six feet. Many members of this family have stinging hairs on the plant, but this plant does not have any. Ramie has been known since ancient times, and the Egyptians wrapped their mummies using ramie. A New World species, *B. cylindrical,* was the source of fiber for the Indians who used it to tie their spearheads and arrowheads to the shafts. The plant is hard to grow for fibers; the extraction is laborious, and the fibers have gum and pectin that make it difficult for processing. Once processed, the fibers are durable, long, and silky and the most desirable of the other soft fibers. Modern techniques in processing have made it easy to extract the fibers. Processers use caustic soda to remove excessive gums that hold the fibers. The United States uses it as a rotation crop and is increasingly cultivating it for use in textiles, mostly mixed with cotton or wool. Ramie fibers are used in making industrial sewing thread, fishing nets, fire hose backings, and lantern mantles. China is the leading producer and exporter of ramie fiber.

Hard Fibers

Hard fibers are extracted from the leaves of certain monocot plants with long parallel veins. The veins are the vascular bundles that run through the entire length of the leaves. Because of this, we also know them as **leaf fibers**. Manufacturers extract the leaf fibers by the decortication method. Such fibers are stiff and unsuitable for textile use, but one can spin them for various uses. There are two main sources: sisal (and henequen), and abaca, both monocots.

Fig. 11.7a Agave plants
Courtesy Rani Vajravelu

Fig. 11.7b Banana plants
Courtesy Rani Vajravelu

Sisal

Sisal and henequen are extracted from the huge, fleshy leaves of *Agave* spp. Agavaceae, which is native to Central America. The plants thrive well in arid, dry regions that are unsuitable for the cultivation of other crops. The uses of fibers were known since the ancient days of the Mayan and Aztec civilizations. They used the sharp spines at the leaf tips and the fiber from the leaves to sew clothing together, hence the name "needle and thread plant." Rough fibers are used for outdoor carpets, mats, sacks, and in tea bags. *Agave sisalana* that yields sisal fibers is planted extensively in Brazil and Africa, which lead in world production. Chapter 6 discusses the use of these plants in alcoholic beverages.

Abaca

Abaca, or Manila hemp, is extracted from the succulent, leaf petiole sheaths of *Musa textilis,* Musaceae. *Musa* is the genus of commercial bananas. Chapter 4 discusses this plant in detail. Abaca is indigenous to the Philippines where the people made clothing from the fibers. The petiole sheaths partially wrap around a central core and end in a crown of tightly packed expanded leaf blades. Before extracting the fiber, workers remove the leaf blades and strip the petiole sheath for fiber. Depending on the height of the banana plant, a fiber can reach up to ten feet long. Abaca fibers are resistant to water decay. The soft brown or ivory white fibers are strong and have a multitude of uses ranging from making manila envelopes, high grade decorative paper, dollar bills, tea bags, coffee filters, vacuum cleaner bags, and food wrappings to filter-tipped cigarettes.

Wood Fiber and Papermaking

The secondary xylem of dicot wood and conifer wood are excellent sources of wood fiber. Manufacturers especially prefer conifer woods because of their elongated xylem tracheids compared to the shorter xylem vessels found in dicots. The **papermaking** industry uses wood fibers.

Different civilizations had invented paper independently from various plant sources. The word paper itself derives from the name of the papyrus plant, *Cyperus papyrus,* Cyperaceae. The Egyptians were the first to use the sheets of papyrus for writing. In the Orient, they used rice paper obtained from *Tetrapanax papyriferus,* Araliaceae as a writing medium. Documents from CE 105 show that first true paper made in China was from hemp. Mayans and Polynesians independently used paper mulberry, *Broussonetia papyrifera,* Moraceae, to make paper like sheets.

The modern papermaking process uses many different fiber sources. Besides wood, other sources include abaca, sugarcane pressings, and bamboo. Even though the source of fiber may differ, the process of papermaking is essentially the same for all.

The Europeans standardized papermaking techniques in the late 1800s. The French invented an earlier papermaking machine in 1789. Before the invention of the machines, people did papermaking by hand. The process was extremely laborious and slow. The resulting paper was dull with poor quality. In spite of this, paper was very expensive. The Europeans used bast fibers from linen and hemp, and also from cotton. Manufacturers make rag bond paper from fibers extracted from rags.

In the case of wood, the first process is breaking the debarked wood into small pieces called **wood chips**. The cheapest paper used in newspapers and catalogs and the least expensive paper towels use the method of mechanically grinding the wood chips into pulp. This pulp contains lignin and pectin that would yellow the paper eventually.

Chemical papermaking processes make the wood pulp by grinding the wood chips using chemicals that dissolve the pectin and lignin that are the components of xylem. This pulp contains fibers, lignin, pectin, resin, and remnants of some bark. Since the desired product is only the fibers, this slurry pulp undergoes washing, cleaning, beating, and bleaching, followed by floating it in a thin sheet on a screen. The workers immediately drain off the water and pass the paper sheet through a series of rollers to press it even and dry. To fill the irregularities of paper and improve its quality, manufacturers add some sizing agents, such as resin, gum, starch, and rosin, before the pulp flows through the screen. This helps the paper to accept the ink when using the paper for printing.

The addition of chemical agents in wood chips determines the quality of the paper. One such chemical used to be the caustic soda (sodium hydroxide) process, which manufacturers use least today. Current processes use the sulfate process that uses a mixture of sodium hydroxide and sodium sulfide. This alkaline process yields "acid free" papers that do not disintegrate prematurely.

The United States is the major consumer of paper and paper products. In recent years, due to increased awareness, society has taken several measures to save the trees and the environment. One such measure is

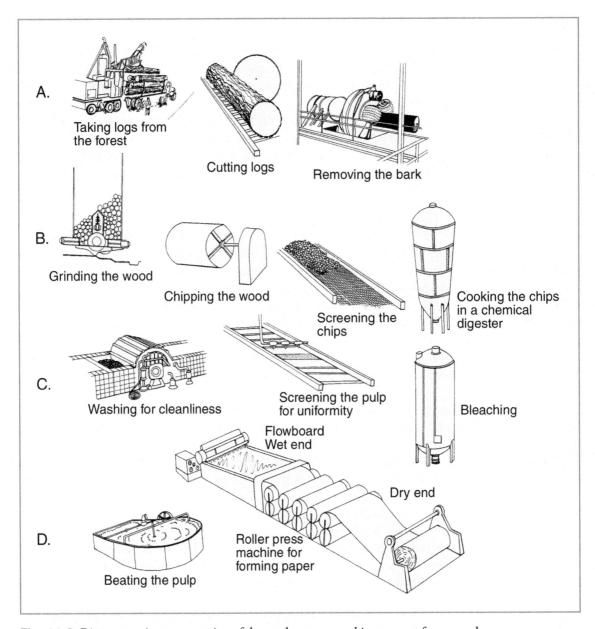

A.
Taking logs from
the forest

Cutting logs

Removing the bark

B.
Grinding the wood

Chipping the wood

Screening the
chips

Cooking the chips
in a chemical
digester

C.
Washing for cleanliness

Screening the pulp
for uniformity

Bleaching

Flowboard
Wet end

Dry end

D.
Beating the pulp

Roller press
machine for
forming paper

Fig. 11.8 Diagrammatic representation of the modern papermaking process from wood.

Row A: Wood logs are cut and debarked.

Row B: Wood is chipped, screened, and cooked in a chemical digester.

Row C: The wood pulp is washed, screened, and bleached.

Row D: The wood pulp is beaten and poured onto a flowboard. Once the excess water is drained off, the pulp moves through a series of rollers. The pressed paper sheet is later dried.

recycling. Manufacturers make recycled paper by redissolving the used paper into slurry, separating the ink, and remaking the paper. Compared to a few years ago, recent technologies have led to mixing all types of paper together for recycling. To improve the quality of recycled paper, they mix fresh wood pulp called virgin pulp with recycled paper slurry. Newly developed computer softwares play a role in controlling the process.

Besides cutting down on the use of disposable paper products, the use of fewer chemicals in the paper industry would make this process environmentally friendly. Another breakthrough is growing trees with less lignin that hopefully will cut down on the amount of chemicals used in papermaking. Genetic engineering has already developed aspen trees, *Populus* spp., Salicaceae, with less lignin and more cellulose.

Wood

Wood is accumulations of secondary xylem. In Greek, *xylon* means wood, from which we derive the word xylem. The meristematic activity of vascular cambium produces secondary xylem. Vascular bundles of woody dicots and gymnosperms develop vascular cambium that in three-dimensional view appear as a continuous cylinder. In the early stages of development, the primary tissues of the stems of young herbaceous dicots, woody dicots, and cone-bearing gymnosperms (conifers) are all in a similar arrangement. In woody plants, however, obvious differences begin to appear as soon as the vascular cambium and the cork cambium develop. As the vascular cambial cells divide, they produce rings of secondary xylem (wood) to the inside and secondary phloem (part of bark) to the outside. The bark and wood increase the plant's girth, thereby making the plant sturdier and wider. Each year, the vascular cambium adds secondary xylem to the inside and secondary phloem to the outside of the vascular cambium. This produces more secondary xylem compared to secondary phloem. As the stems and roots grow in thickness, older layers of secondary vascular tissues along with mature primary tissues become pushed farther apart. Older phloem becomes fiber cells and intermingles with bark tissues to make up the **bark** of the tree. The sturdier xylem cells keep accumulating in distinct rings, each ring comprised of the light-colored **spring wood** or early wood, followed by the dark-colored **summer wood** or late wood. Botanists call these rings **annual growth rings**. You can estimate the age of a temperate tree by counting the number of annual rings; in tropical trees such rings are not readily visible due to the lack of distinct seasons.

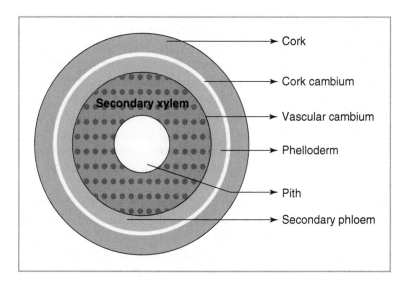

Fig. 11.9 A diagrammatic view of a woody stem in cross section
Courtesy Rani Vajravelu

Characteristics of Wood

Timber or lumber is wood, an important product of woody trees. Manufacturers supply lumber in an unfinished form called rough lumber or in a finished form cut from logs into standard-size boards. An air drying process called seasoning removes the water content of the wood. They grade and sort seasoned lumber based on the tree type it was cut from originally. Based on its overall figure (described below), buyers choose lumber for different purposes, including construction and furniture making.

The lumber industry distinguishes two kinds of woods based on the plants that produce the wood and the basic components of the wood tissue. Wood produced by gymnosperms (especially conifers) consists mainly of xylem tracheids and fibers, but it lacks the xylem vessels. Botanists consider this wood **soft wood**. Woods of dicots have xylem vessels along with tracheids and xylem fibers known as **hard wood**. The terms soft and hard do not relate to the hardness of the wood itself, but rather to the plants that produced it.

In the wood industry, sellers price and choose wood based on its color, porosity, grain, and figure, all of which contribute to its appearance. Based on the impregnation on the xylem, the **color** of the wood may vary from light brown to yellow, red, purple, or to dark brown. Builders prefer each color in a certain type of furniture piece or in musical instruments.

Porocity refers to the size and distribution of large xylem vessels through the wood. Ring porous wood has xylem vessels arranged in a ring; diffuse porous wood have them dispersed throughout. The alignment of xylem vessels determines the **grain** of the wood as either straight or cross grained. The latter type is hard to work with and produces more sawdust compared to the straight grain type. The **figure** of wood combines all of the above features in a wood that determines the structural quality of wood. The wood industry considers the figure of wood as an important basis of selection for various uses.

Any wood that has a low **density** (mass in grams divided by volume) like balsa that has a value of <1 is considered to have a light density. Lignum vitae has a density of >1, which makes it hard for carpenters to work with. They usually prefer oak wood, with a medium density of around 0.60 g/cm3, for furniture making. Home construction uses pine woods that have a density of about 0.35 to 0.50. Maples and Indian teak woods have a density of 0.62–0.80, and furniture makers use them in making quality furniture pieces.

Wood Products

Veneers are thin uniform slices of desirable wood glued onto another surface to create attractive wood pieces. Manufacturers make veneers by slicing large rectangular blocks or by peeling a thin piece of stem (less than 5 mm diameter) and continuously shaving the revolving log using a rotary machine. Fruit crates use rough veneers, and finer pieces cover musical instruments, wood furniture, particle board, or medium density fiberboards used in doors, cabinets, furniture, and so forth. Using veneer-covered pieces is cheaper and less effort than making fine wood pieces from real hard wood and doesn't sacrifice the look.

Plywood is made from veneer sheets glued together at right angles to the adjacent layers for strength. Normally manufacturers attach three or more pieces together, thus creating three-ply, five-ply sheets, always with an odd number. They also make plywoods with a solid core covered by an even number of veneers on each side. Some carefully crafted plywoods may be stronger than most solid wood boards or even steel as long as they use quality veneers.

Ancient Egyptian and Roman times used veneers and plywoods. Egyptian tombs dated 1500 BCE show wood pieces very similar to modern veneers. Marquetry, a wood craft using inlaid pieces of veneers, was very popular during the time of King Louis XIV of France.

Particle boards are relatively recent in wood history. Their creation began in the 1940s as a replacement for plywood. Particle boards make use of undesirable leftover wood pieces, and they are cheaper than conventional wood and plywood. Manufacturers use resin to glue together wood chips, saw mill shavings, and sawdust into a composite board. They glue veneers onto their surfaces to make them desirable. They also make particleboards from recycled wood, which cuts down on the use of fresh virgin wood from trees. Unlike solid wood that is prone to warping, particle boards are stable, enabling their use in new design possibilities.

Fiberboards are made in a similar manner except they use the wood fibers. Manufacturers make the wood fibers either by mechanical or by chemical processing in a manner very similar to the wood fiber slurry used in papermaking. They add various protective compounds for fire and pest resistance along with resin that determines its strength. The furniture industry uses medium-density fiberboards.

In addition to the above products, the industry cuts and shreds the bark of woody trees for mulches. They dissolve the cellulose in wood and use it in making cellophane and rayon, a semisynthetic viscose fiber. Cellophane is chemically identical to rayon but made into sheets instead of fibers.

Woody Stems

Teak, *Tectona grandis*, is a tropical hardwood deciduous tree from Verbenaceae. It is native to southeast Asia and has a wide distribution in India. The teak wood is weather resistant. Because of this it is preferable in making outdoor furniture, boat decks, doors and window frames, and flooring and wooden columns.

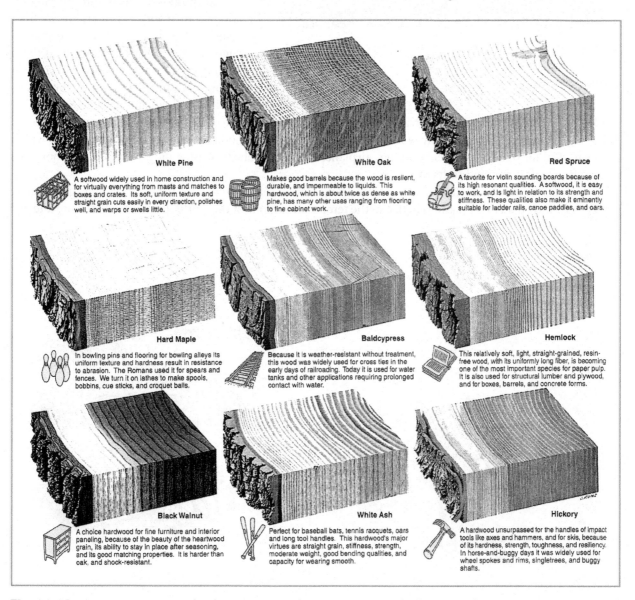

White Pine
A softwood widely used in home construction and for virtually everything from masts and matches to boxes and crates. Its soft, uniform texture and straight grain cuts easily in every direction, polishes well, and warps or swells little.

White Oak
Makes good barrels because the wood is resilient, durable, and impermeable to liquids. This hardwood, which is about twice as dense as white pine, has many other uses ranging from flooring to fine cabinet work.

Red Spruce
A favorite for violin sounding boards because of its high resonant qualities. A softwood, it is easy to work, and is light in relation to its strength and stiffness. These qualities also make it eminently suitable for ladder rails, canoe paddles, and oars.

Hard Maple
In bowling pins and flooring for bowling alleys its uniform texture and hardness result in resistance to abrasion. The Romans used it for spears and fences. We turn it on lathes to make spools, bobbins, cue sticks, and croquet balls.

Baldcypress
Because it is weather-resistant without treatment, this wood was widely used for cross ties in the early days of railroading. Today it is used for water tanks and other applications requiring prolonged contact with water.

Hemlock
This relatively soft, light, straight-grained, resin-free wood, with its uniformly long fiber, is becoming one of the most important species for paper pulp. It is also used for structural lumber and plywood, and for boxes, barrels, and concrete forms.

Black Walnut
A choice hardwood for fine furniture and interior paneling, because of the beauty of the heartwood grain, its ability to stay in place after seasoning, and its good matching properties. It is harder than oak, and shock-resistant.

White Ash
Perfect for baseball bats, tennis racquets, oars and long tool handles. This hardwood's major virtues are straight grain, stiffness, strength, moderate weight, good bending qualities, and capacity for wearing smooth.

Hickory
A hardwood unsurpassed for the handles of impact tools like axes and hammers, and for skis, because of its hardness, strength, toughness, and resiliency. In horse-and-buggy days it was widely used for wheel spokes and rims, singletrees, and buggy shafts.

Fig. 11.10 A representative sample of angiosperm and gymnosperm woods illustrating their characteristics and specific uses.

Fig. 11.11a Sandalwood
© niyoseris/Shutterstock.com

Maple is a common name for about 120 species of the genus *Acer* included in Sapindaceae. The trees are medium sized, deciduous or evergreen, and common in temperate regions of Europe, Asia, North America, and Africa. Manufacturers use maple wood in making percussion instruments like drum kits. They use sugar maple wood in making bowling pins, baseball bats, and musical instruments. One species, *Acer palmatum,* has many cultivars, mainly grown in Japan.

Mahogany, *Swietenia mahagoni,* Meliaceae wood is preferable in boat making, furniture, and musical instruments, especially drums and guitars. Mahogany wood has a beautiful reddish color that shines when polished. It is straight grained and durable.

Balsa, *Ochroma pyramidale,* Malvaceae, is a deciduous tree native to tropical South and Central America. Balsa has the lowest density for any wood and is the lightest hardwood. The timber is soft yet strong, which makes it ideal for building model buildings, airplanes, bridges, lifebelts, fishing plugs, surfboards, floor pans in cars, fine craft items, and so forth.

Oak is a common name that refers to a large group of trees that belong to the genus *Quercus,* Fagaceae. The genus has a wide distribution and is native to the temperate regions of the world. It includes the white and red

Fig. 11.11b Casuarina
© Dennis Tokarzewski/Shutterstock.com

Fig. 11.11c Rice paper plant
© Morphart Creation/Shutterstock.com

oaks *Q. alba* and *Q. rubra,* respectively. Due to the wood's ideal density and strength, the wood industry much favors oak. Due also to the presence of tannin, the wood is naturally resistant to pest attack. The attractive grain of oak boards has made them popular since ancient times for the interior paneling of prestigious buildings. In Europe, shipbuilders constructed ships using oak wood. The use of oak wood still continues in flooring, furniture, paneling, and veneer making. Oak barrels store fermented alcoholic beverages such as wine and brandy as they age. The bark of *Quercus sober* or cork oak is the source of natural cork, used in bottle stoppers, shoe soles, fishing floats, and as spaceship insulation.

Pines are cone bearing trees of the gymnosperm group. There are about a hundred or more species of the genus *Pinus,* Pinaceae. The trees are tall and evergreen with needle-like leaves. Pine trees are native to the northern hemisphere but also thrive well in subtropical to temperate regions of the world. Pine wood is classified as softwood, even though it is much stronger than most angiosperm hard woods. Throughout the world, people value pines for their timber and wood pulp. Long, cylindrical, unbranched tree trunks are used in telephone poles. Sticky, aromatic resin is an important source of turpentine.

Bamboos are a large group of distinct grasses classified in a separate subfamily Bambusoideae, Poaceae. There are both woody and herbaceous bamboos. Distribution of the woody bamboos is in tropical and temperate regions. Bamboos are the fastest growing plants in the world. The stems do not undergo secondary growth. The woody stems develop by the primary thickening of the tissues formed by the apical meristem. Most bamboos have hollow internodes (culm) and solid nodal regions. When treated,

TABLE 11.1 Other sources of fiber and wood

Common Name	Binomial and Family	Distribution	Uses
Alpine ash	*Eucalyptus gigantea Myrtaceae*	Southeastern Australia, southeast Asia	Pale brown wood, straight grained; flooring, cabinetry
Casuarina	*Casuarina equisetifolia Casuarinaceae*	Australia, Pacific Islands; southeast Asia	Heavy, fine-grained thin, cylindrical wood used in carpentry, bullock-yokes, wheel spokes, cottages
Blue gum	*Eucalyptus globulus Myrtaceae*	Australia; cultivated elsewhere	Pale, heavy wood, used in ship building, wood pulp. Oil extracted, used in medicine
Giant reed	*Arundo donax*	Mediterranean; tropical Asia	Mats, lattices, screens
Kenaf	*Hibiscus cannabinus Malvaceae*	India; Africa	Yields coarse bast fibers similar to jute
Pineapple	*Ananas comosus Bromeliaceae*	Central America	Piña fibers in shirts, scarves; extracted from leaves
Rattan	*Calamus gibbsianus and several other genera Arecaceae*	Borneo Rattan palms are all over tropical Asia, Australia, and Africa.	Furniture, canes, handles of musical instruments
Rice paper plant	*Tetrapanax papyriferus Araliaceae*	Taiwan; Southern China	Fiber from pith used in making rice paper
Sandalwood	*Santalum lanceolatum Santalaceae*	Australia, Pacific Islands; southeast Asia	Fragrant wood in craft items; selectively in furniture

bamboo forms a very hard wood that is lightweight and extremely durable. Consumers use whole bamboo stems in building construction, bridges, fences, canoes, fishing rods, chopsticks, and dinner mats. They also use stem shavings in dinnerware, furniture, flooring, musical instruments, knitting needles, and many more products. Bamboos yield paper pulp, and in some genera people eat young shoots as a delicacy.

Table 11.2 lists the important genera and their main uses and distribution.

TABLE 11.2 Important bamboos around the world

Binomial	Plant Features	Distribution	Uses
Arundinaria amabilis	Strong and flexible, culms about 12 m tall	China—highly priced	Fishing rods, ski poles, furniture
Bambusa vulgaris	Very strong culms; up to 20 m tall	Known only in cultivation	Paper pulp; young shoots edible
Dendrocalamus strictus	Some forms with solid internodes; culms up to 15 m tall	Most common, universally distributed in India	Paper pulp
Dendrocalamus asper	Culms about 7 m tall, clumping type	Cultivated in Malaysia and Indonesia	Ornamental; musical instruments; large edible shoots
Dendrocalamus giganteus	Giant bamboo; culms over 30 m tall	Sri Lanka	Building purposes; weaving
Gigantochloa apus	Culms are 10–20 m tall; 9 cm diameter	Most common in West Java	Building purposes; edible shoots; paper pulp
Guadua angustifoila	Very strong, durable, culms about 20–30 m tall	Common in North and South America	Building purposes; poles for cables, bridges, water pipes; flooring
Melocanna baccifera	Culms up to 20 m tall	Bangladesh and Burma (Myanmar)	Building purposes; basketry, matting; paper pulp
Oxytenanthera abyssinica	Solid internodes; culms up to 15 m tall	Tropical Africa	Building purposes; grains eaten rarely

Review Questions

Give brief answers in the given space.

1. Define the word "fiber."

2. Outline the methods of extracting bast fibers.

3. What is coir? Give the name and family of the plant that gives us coir.

4. How many major types of cotton are in the world? Give the common names and their distribution.

5. Identify at least five plant sources of commercial fiber that we have already discussed in previous chapters.

6. List three ancient sources of plant fibers that were in use prior to wood pulp.

7. What is the "figure" of a wood?

8. How do manufacturers make and use fiberboards?

9. Give the binomials of two bamboos that give us paper pulp.

10. Discuss future directions for the paper industry that would be more environmentally friendly.

Multiple Choice Questions

For each question, choose one best answer.

1. When a New York importer first prepared tea bags, he used silk for a bag. Present day tea bags might contain which fiber?
    A. Coir
    B. Abaca
    C. Sisal
    D. All of the above
    E. Only B and C

Match the following fibers from Column 1 to the appropriate sources from Column 2. You may use a choice once, more than once, or not at all.

Column 1	Column 2
2. Flax	A. *Gossypium*
	B. *Corchorus*
3. Ramie	C. *Linum*
4. emp	D. *Cannabis*
	E. *Boehmeria*

5. Coir fibers come from of _____ of _____.
    A. endocarp; kapok
    B. seeds; cotton
    C. mesocarp; coconut
    D. exocarp; coconut
    E. only A and D

6. The term for the microbial degradation of soft tissues to free the fibers in the processing of linen from Linum is _____.

   A. skutching
   B. carding
   C. combing
   D. retting
   E. decorticating

7. Plywood is:

   A. thin sheets of desirable wood
   B. veneers glued together at right angles to each other
   C. wood chips glued together into a board
   D. sawdust glued together onto a board
   E. solid wood pieces with a standard size

8. Balsa wood is:

   A. a dicot
   B. a hardwood
   C. a softwood
   D. both A and B
   E. both A and C

9. Which of the following would be desirable in paper making industry?

   A. teak
   B. bamboo
   C. mahogany
   D. pine
   E. maple

10. Besides paper and timber, manufacturers also use wood in making:

   A. bark
   B. rayon
   C. cellophane
   D. all of the above
   E. only B and C

True/false Questions

1. Industry uses durable fibers of cotton in making fire hose backing threads.   **True   False**

2. Cellulose determines the tensile strength of a fiber.   **True   False**

3. The mesocarp of coconut provides bast fibers.   **True   False**

4. Cotton fibers have more cellulose than coir fibers.   **True   False**

5. Plant fibers used in dollar bills come from bleached flax.   **True   False**

6. Bamboo is unique in that it is the only monocot that undergoes secondary growth   **True   False**

7. The Egyptians standardized papermaking techniques in 1800s.   **True   False**

8. We also refer to conifer woods as softwood.   **True   False**

9. Wood is accumulations of secondary xylem.   **True   False**

10. Turpentine is a product of pine trees.   **True   False**

APPENDIX 1

Leading producers of the plants mentioned in this book.

Vegetables: Leading producers (as of 2012)			
Common Name	Leading Producer	Secondary Producer	Tertiary Producer
Artichoke	Egypt	Italy	Spain
Asparagus	China	Peru	Mexico
Beets (sugar)	Russian Federation	France	USA
Brussel sprouts*	China	India	Russian Federation
Cabbage*	China	India	Russian Federation
Carrot**	China	Russian Federation	USA
Cassava	Nigeria	Indonesia	Thailand
Cauliflower and Broccoli	China	India	Italy
Chicory root	Belgium	France	Poland
Chinese Cabbage*	China	India	Russian Federation
Collard greens*	China	India	Russian Federation
Ginger	India	China	Nepal
Kohlrabi*	China	India	Russian Federation
Lettuce	China	USA	India
Onion	China	India	USA
Savoy Cabbage*	China	India	Russian Federation
Spinach	China	USA	Japan
Sweet potato	China	Nigeria	Uganda
Taro	Nigeria	China	Ghana
Turnip**	China	Russian Federation	USA
White potato	China	India	USA
Yam	Nigeria	Ghana	Cote d'Ivoire

*Cabbage and other brassicas (together)
**Carrots and turnips (together)

(Continued)

Fruits: Leading producers (as of 2012)			
Common Name	Leading Producer	Secondary Producer	Tertiary Producer
Almond	USA	USA	Spain
Apple	China	USA	Turkey
Apricot	Turkey	Iran (Islamic Republic of)	Uzbekistan
Avocado	Mexico	Indonesia	Dominican Republic
Banana	India	China	Philippines
Blackberry			
Blueberry	USA	Canada	Poland
Brazil nut	Bolivia (Plurinational State of)	Brazil	Cote d'Ivoire
Cashew nut	Vietnam	Nigeria	India
Cashew apple	Brazil	Mali	Madagascar
Cherry (sour)	Turkey	Russian Federation	Poland
Cherry (sweet)	Turkey	USA	Iran (Islamic Republic of)
Chestnut	China	Republic of Korea	Turkey
Chilis and Peppers	China	Mexico	Turkey
Coconut	Indonesia	Philippines	India
Corn (maize)	USA	China	Brazil
Cranberry	USA	Canada	Belarus
Cucumbers and gherkins	China	Turkey	Iran (Islamic Republic of)
Custard apple			
Dates	Egypt	Iran (Islamic Republic of)	Saudi Arabia
Eggplant	China	India	Iran (Islamic Republic of)
Fig	Turkey	Egypt	Algeria
Grapefruit and pomelos	China	USA	Mexico
Grapes	China	USA	Italy
Guava*	India	China	Kenya
Hazelnut	Turkey	Italy	USA
Kiwi	Italy	New Zealand	Chile
Lemons and Limes	China	India	Mexico
Mango*	India	China	Kenya
Mangosteen*	India	China	Kenya
Melons	China	Turkey	Iran (Islamic Republic of)
Okra	India	Nigeria	Iraq
Olive	Spain	Italy	Greece
Orange (sweet)	Brazil	USA	China

Fruits: Leading producers (as of 2012) (Continued)		
Common Name	**Leading Producer**	**Secondary Producer**	**Tertiary Producer**
Papaya	India	Brazil	Indonesia
Peach and nectarine	China	Italy	USA
Pear	China	USA	Argentina
Pineapple	Thailand	Costa Rica	Brazil
Pistachios	Iran (Islamic Republic of)	USA	Turkey
Plum	China	Romania	Serbia
Pumpkin, squash	China	India	Russian Federation
Quince	Turkey	China	Uzbekistan
Raspberry	Russian Federation	Poland	USA
Strawberry	USA	Mexico	Turkey
Tangerine	China	Spain	Brazil
Tomato	China	India	USA
Walnut	China	Iran (Islamic Republic of)	USA

Grains and Legumes: Leading producers (as of 2012)		
Common Name	**Leading Producer**	**Secondary Producer**	**Tertiary Producer**
Barley	Russian Federation	France	Germany
*Buckwheat	Russian Federation	China	Ukraine
Corn (maize)	USA	China	Brazil
Millet	India	Nigeria	Niger
Oats	Russian Federation	Canada	Poland
*Quinoa	Bolivia (Plurinational State of)	Peru	Ecuador
Rice	China	India	Indonesia
Rye	Germany	Poland	Russian Federation
Sorghum	Nigeria	USA	India
Sugar cane	Brazil	India	China
Triticale	France	Germany	Belarus
Wheat	China	India	USA

*Not true grains

(Continued)

Grains and Legumes: Leading producers (as of 2012) Continued			
Common Name	Leading Producer	Secondary Producer	Tertiary Producer
Beans, green	China	Indonesia	India
Black-eyed peas	Nigeria	Niger	Burkina Faso
Broad beans	China	Ethiopia	Australia
Carob	Spain	Portugal	Greece
Chick-peas	India	Australia	Turkey
Lentils	Canada	India	Australia
Peanut	China	India	Nigeria
Peas, green	China	India	France
Pigeon peas	India	Myanmar	Malawi
Soybean	USA	Brazil	Argentina
String beans	USA	France	Morocco

Beverage sources: Leading producers (as of 2012)				
Common Name	Leading Producer	Secondary Producer	Tertiary Producer	Leading from book
Agave	Colombia	Mexico	Nicaragua	
Asi	USA			
Barley	Russian Federation	France	Germany	
Cacao	Cote d'Ivoire	Indonesia	Ghana	Africa
Chicory	Belgium	France	Poland	
Coffee	Brazil	Viet Nam	Indonesia	Brazil
Cola	Nigeria	Cote d'Ivoire	Cameroon	Africa
Corn	USA	China	Brazil	
Grapes	China	USA	Italy	
Guarana	Brazil			
Guayusa	Ecuador	Peru/Colombia		
Hops	Germany	USA	Ethiopia	
Mate	Brazil	Argentina	Paraguay	
Rice	China	India	Indonesia	
Sugarcane	Brazil	India	China	
Tea	China	India	Kenya	India

Vegetable Oils: Leading producers of the plants mentioned (as of 2012)			
Common Name	Leading Producer	Secondary Producer	Tertiary Producer
Bayberry	southeast USA, Carribean, Central America		
Canola	Canada	China	India
Carnauba wax	Brazil		
Castor bean	India	China	Mozambique
Cottonseed plant	China	India	USA
Flax	Canada	Russian Federation	China
Neem	India	Asia, Africa, Middle East	
Palm kernal	Indonesia	Malaysia	Thailand
Rice Bran	India	China	
Safflower	Mexico	India	Kazakhstan
Sesame	Myanmar	India	China
Sunflower	Ukraine	Russian Federation	Argentina
Tung	China	Paraguay	Argentina

Note: Plants mentioned elsewhere are omitted here, unless the leading producer for oil is different from another product for a plant. Ex. Rice (as grain or as oil source).

Spices and Herbs: Leading producers (as of 2012)			
Common Name	Leading Producer(s)	Secondary Producer	Tertiary Producer
Ajwain	India, Iran		
Allspice	Jamaica	Honduras, Guatemala, Leewards Islands	
Anise	India	Mexico	China
Asafoetida	India		
Basil	USA	European countries and in India	
Capers	Morocco, Spain, Turkey, Italy, Aeolian Islands		
Caraway	Netherlands	European countries, USA	
Cardamom	Guatamala	Indonesia	India
Chili pepper	India	China	Peru
Cinnamon	Indonesia	China	Viet Nam
Cloves	Indonesia	Madagascar	United Republic of Tanzania
Coriander	India	Mexico	China
Cumin	India	Syria, Turkey, Iran, China	

(Continued)

Spices and Herbs: Leading producers (as of 2012)			
Common Name	Leading Producer(s)	Secondary Producer	Tertiary Producer
Dill	India, Pakistan	European countries, USA	
Fennel	India	Mexico	China
Fenugreek	India	European and South American countries, USA	
Ginger	India	China	Nepal
Horseradish			
Indian curry leaf	India	Sri Lanka	
Mace	Guatamala	Indonesia	India
Mango ginger	India		
Marjoram	European and South American countries, USA		
Mustard, black	Nepal	Canada	Myanmar
Mustard, brown	Nepal	Canada	Myanmar
Mustard, white	Nepal	Canada	Myanmar
Nutmeg	Guatamala	Indonesia	India
Oregano	Turkey	Albania, France, Greece, Italy, Mexico, Spain, Yugoslavia	
Parsley	European countries, Canada & Japan		
Pepper	Viet Nam	Indonesia	India
Peppermint	Morocco	Argentina	Spain
Rosemary	European and South American countries, USA		
Saffron	Iran	Greece	Morocco, India, Italy, Spain
Sage	European and South American countries, USA		
Spearmint	Germany, Japan, Netherlands, Russia		
Star anise	India	Mexico	China
Tarragon	Russia, France, USA		
Thyme	Spain	European countries, Canada, USA	
Turmeric	India	Pakistan, China, Haiti, Jamaica, Peru, Taiwan, Thailand	
Vanilla	Madagascar	Indonesia	China
Wasabi	Japan		

Medicinal Plants: Leading producers (as of 2012)			
Common Name	**Leading Producer(s)**	**Secondary Producer**	**Tertiary Producer**
Aloe	Central America, southern USA		
Belladonna	Western Europe, India, Pakistan	Honduras, Guatemala, Leewards Islands	
Coca	Colombia	Peru	Bolivia
Ephedra	China		
Fever bark tree	India, Indonesia, Zaire, Tanzania, Kenya, Sri Lanka, Bolivia, Colombia, Costa Rica	European countries and in India	
Foxglove	USA, UK, Netherlands, Switzerland, Germany, former USSR		
Henbane	USA, UK, India	European countries, USA	
Marijuana	Afghanistan	Jamaica, Mexico, Paraguay	India, South Africa, Canada, Colombia, Congo, Thailand
Mayapple	Eastern North America	China	Peru
Meadow saffron	Europe, North Africa	China	Viet Nam
Opium poppy	Afghanistan, India (licit)	Mexico	Burma
Pacific yew	North America (coastal)	Mexico	China
Papaya	India	Brazil	Indonesia
Periwinkle	India, Madagascar, Israel, USA		
Snakeroot	India, Thailand, Bangladesh, Sri Lanka	western coast of Africa: Zaire, Mozambique, Rwanda	
White Willow	eastern and southeastern Europe		
Wormwood	China, Viet Nam, Africa, India		

Plant Narcotics and Poisons: Leading producers (as of 2012)			
Common Name	**Leading Producer(s)**	**Secondary Producer**	**Tertiary Producer**
Angel's trumpet	South America (tropical), USA (south), Puerto Rico, Pacific islands		
Belladonna	Western Europe, India, Pakistan	Honduras, Guatemala, Leewards Islands	
Betel	India	Bangladesh	Sri Lanka
Caapi	South America, western Amazon regions (Peru, Ecuador, Bolivia)		

(Continued)

Plant Narcotics and Poisons: Leading producers (as of 2012) (Continued)			
Common Name	Leading Producer(s)	Secondary Producer	Tertiary Producer
Castor bean	India	China	Mozambique
Coca	Colombia	Peru	Bolivia
Curare	South America		
Curare	South America		
Dumbcane	South and Central America		
Henbane	USA, UK, India		
Jack-in-the-pulpit	USA, Canada		
Jequirity bean	India	Carribean	
Jimson weed	USA (tropical North America)		
Kat (Khat)	Ethiopia	Somalia, Yemen	
Kava	Fiji, Samoa, Tonga, Vanuata		
Mandrake	Europe: northern Italy, former Yugoslavia		
Marijuana	Afghanistan	Jamaica, Mexico, Paraguay	India, South Africa, Canada
Mescal bean	North America		
Milkweed	North America		
Nux vomica	Asia, West Africa		
Oleander	Mediterranean region, Arabia, Southwest Asia		
Ololiuqui	Mexico		
Ololiuqui	Mexico		
Opium	Afghanistan, India (licit)	Mexico	Burma
Peyote	USA (southwest), Mexico		
Poison hemlock	Europe, North Africa, West Asia, North America, Australia, New Zealand		
Skunk cabbage	USA, Canada		
Tobacco	China	Brazil	India

Plant Fibers and Timber: Leading producers (as of 2012)			
Common Name	Leading Producer (as of 2012)	Secondary Producer	Tertiary Producer
Coconut (plant)	Indonesia	Philippines	India
Cotton (lint)	China	India	USA
Kapok	Indonesia	Thailand	
Jute	India	Bangladesh	China
Linen (flax)	France	Belarus	Russian Federation
Hemp (tow waste)	China	Democratic People's Republic of Korea	Netherlands
Ramie	China	Lao People's Democratic Republic	Philippines
Sisal	Brazil	Kenya	United Republic of Tanzania
Abaca (manila)	Philippines	Ecuador	Costa Rica
Timber	USA	India	China

The following references were used to compile the above data listed under Appendix 1.

http://faostat.fao.org/site/339/default.aspx (2012)

http://articles.economictimes.indiatimes.com/2013-05-25/news/39521452_1_oryzanol-refined-rbo-edible-oil

http://edition.cnn.com/2008/WORLD/americas/11/25/paraguay.mexico.marijuana/ (2008)

http://keys.lucidcentral.org/keys/v3/eafrinet/weeds/key/weeds/Media/Html/Brugmansia_suaveolens_(Angels_Trumpet).htm

http://turmericworld.com/production.php

http://www.faculty.ucr.edu/~legneref/botany/medicine.htm

http://www.fao.org/docrep/x5326e/x5326e0e.htm (1991)

http://www.ncbi.nlm.nih.gov/pmc/articles/

http://www.soyatech.com/

https://en.wikipedia.org/

https://www.afpm.org/wax-facts/

https://www.cia.gov/library/publications/the-world-factbook/fields/2086.html

http://ageconsearch.umn.edu/bitstream/126066/2/Xin.pdf

http://food-foodstuff.com/guarana_9234_0

http://www.jamaicaobserver.com/news/Country-remains-largest-producer--exporter-of-marijuana-in-Caribbean---report-_18502845 (2015)

http://www.mapsofworld.com/world-top-ten/countries-with-most-timber-producing-countries.html

http://www.nda.agric.za/docs/Brochures/ProGuiBasil.pdf

http://www.nydailynews.com/news/world/afghanistan-world-top-marijuana-supplier-article-1.173257 (2010)

http://www.smokingwithstyle.com/which-countries-produce-the-most-cannabis/

https://www.cips.org/Documents/Knowledge/Categories-Commodities/Mintec/MintecCFS_Cumin.pdf

Ben-Erik van Wyk & M. Wink Medicinal Plants of the World, 2009, Timber Press

Peter, K.V. Handbook of Herbs and Spices, Volume 2.2004

Simpson, B.B., and M.C. Ogorzaly. 2001. Economic Botany, 3rd ed. New York: Mc-Graw-Hill Higher Education. Small, Ernest, Culinary Herbs. 2006.

APPENDIX 2

Answers to the Review Questions given at the end of each chapter.

Answers to the first ten questions that require brief answers are of free response type. If you do not know the answers for these questions, please read the appropriate areas of the book which will help you review the material. For this reason, answers to those questions are not given here.

CHAPTER 1 M/C Questions	1	D
	2	C
	3	B
	4	A
	5	B
	6	C
	7	D
	8	E
	9	E
	10	D
True or False	1	F
	2	F
	3	F
	4	T
	5	F
	6	F
	7	F
	8	T
	9	T
	10	T

CHAPTER 2 M/C Questions	1	B
	2	E
	3	C
	4	E
	5	E
	6	B
	7	E
	8	E
	9	A
	10	A
True or False	1	T
	2	F
	3	F
	4	T
	5	F
	6	T
	7	T
	8	T
	9	T
	10	F

CHAPTER 3M/C Questions	1	E
	2	C
	3	C
	4	C
	5	B
	6	D
	7	B
	8	A
	9	B
	10	B
True or False	1	F
	2	T
	3	T
	4	F
	5	F
	6	T
	7	T
	8	T
	9	F
	10	F

CHAPTER 4 M/C Questions	1	E
	2	B
	3	D
	4	A
	5	D
	6	D
	7	D
	8	D
	9	C
	10	D
True or False	1	T
	2	F
	3	T
	4	T
	5	T
	6	F
	7	T
	8	T
	9	F
	10	F

CHAPTER 5 M/C Questions	1	C
	2	C
	3	B
	4	D
	5	B
	6	B
	7	D
	8	B
	9	C
	10	B
True or False	1	F
	2	F
	3	T
	4	F
	5	F
	6	T
	7	T
	8	T
	9	F
	10	T

CHAPTER 6 M/C Questions	1	A
	2	A
	3	B
	4	C
	5	C
	6	D
	7	D
	8	D
	9	C
	10	E
True or False	1	T
	2	F
	3	T
	4	F
	5	F
	6	F
	7	F
	8	F
	9	T
	10	F

CHAPTER 7 M/C Questions	1	B
	2	C
	3	D
	4	E
	5	D
	6	E
	7	B
	8	B
	9	E
	10	C
True or False	1	F
	2	T
	3	T
	4	T
	5	F
	6	F
	7	F
	8	T
	9	T
	10	F

CHAPTER 8 M/C Questions	1	D
	2	C
	3	D
	4	D
	5	D
	6	E
	7	A
	8	B
	9	B
	10	B
True or False	1	F
	2	T
	3	F
	4	T
	5	T
	6	F
	7	F
	8	F
	9	T
	10	T

CHAPTER 9 M/C Questions	1	D
	2	C
	3	B
	4	C
	5	C
	6	D
	7	B
	8	C
	9	B
	10	C
True or False	1	F
	2	T
	3	F
	4	T
	5	F
	6	F
	7	T
	8	T
	9	T
	10	F

CHAPTER 10 M/C Questions	1	D
	2	B
	3	C
	4	D
	5	C
	6	D
	7	A
	8	B
	9	B
	10	C
True or False	1	T
	2	F
	3	F
	4	T
	5	T
	6	T
	7	F
	8	F
	9	T
	10	F

CHAPTER 11 M/C Questions	1	B
	2	C
	3	E
	4	D
	5	C
	6	D
	7	B
	8	D
	9	D
	10	E
True or False	1	F
	2	T
	3	F
	4	T
	5	F
	6	F
	7	F
	8	T
	9	T
	10	T

REFERENCES

Allen, B. M. 1967. *Malayan Fruits.* Singapore: Donald Moore Press Ltd.

Anderson, E.N., Pearsall, D., Hunn E., and Turner N. *Ethnobiology* edited book 2011 Wiley-Blackwell. Published by John Wiley & Sons, Inc.

Armstong, W.P. 2001. Wayne's Word: 9 May 2001. http://waynesword.palomar.edu/index.htm (21 January 2015).

Ben-Erik van Wyk & M. Wink. Medicinal Plants of the World, 2009, TimberPress.

Coon, Nelson. 1974. *The Dictionary of Useful Plants.* Emmaus, PA: Rodale Press.

Gerard, J. 1975. The Herbal or General History of Plants, Dover Publications, Inc.

Gledhill, D. 2008. *The Names of Plants*, 4th ed. Cambridge, U.K.: Cambridge University Press.

http://www.accessexcellence.org

http://www.agmrc.org

http://www.ars-grin.gov

http://www.botgard.ucla.edu

http://www.cieer.org

http://www.corn.org

http://www.entheology.org

http://www.ico.org

http://www.libraryindex.com

http://www.missouribotanicalgarden.org

http://www.nal.usda.gov/fnic

www.kew.org/science-conservation/collections/herbarium

www.opensiuc.lib.siu.edu~ebl

http://www.tea.co.uk

Judd, W. S., C. S. Campbell, E. A. Kellogg, P. F. Stevens, and M. J. Donoghue. 2008. *Plant Systematics*, 3rd ed. Sunderland, MA: Sinauer Associates, Inc.

Levetin, E., and K. McMahon. 2008. *Plants and Society.* New York: McGraw-Hill Higher Education.

Matthew, K. M. 1983. *The Flora of the Tamilnadu Carnatic.* Tiruchirapalli, India: Diocesan Press.

Peter, K.V. *Handbook of Herbs and Spices*, Volume 2.2004

Purseglove, J. W. 1982. *Tropical Crops Dicotyledons.* Essex, U.K.: Longman Group Ltd.

Purseglove, J. W. 1985. *Tropical Crops Monocotyledons.* New York: Longman Inc.

Schery, Robert W. 1972. *Plants for Man*, 2nd ed. Prentice-Hall, Inc.

Schultes, R. E., and A. Hoffmann. 1979. *Plants of the Gods.* McGraw-Hill Book Company, (UK) Ltd.

Simpson, B. B., and M. C. Ogorzaly. 2001. *Economic Botany*, 3rd. ed. New York: McGraw-Hill Higher Education.

Small, Ernest, *Culinary Herbs.* 2006.

Swerdlow, Joel L. 2000. *Nature's Medicine: Plants That Heal,* National Geographic Society, Washington, D.C.

Uphof, J. C. Th. 1968. *Dictionary of Economic Plants,* 2nd ed. New York: Verlag von J. Cramer, Stechert-Hafner Service Agency, Inc.

Wexler, Barbara 2007. *Antioxidants,* Woodland Publishing, UT

Whitney, E. N., and S. R. Rolfes. 2008. *Understanding Nutrition*, 11th ed. Florence, KY: Cengage Learning.

Wunderlin, R. P., and B. F. Hansen. 2003. *Guide to the Vascular Plants of Florida*, 2nd ed. Gainesville, FL: University Press of Florida.

Yanaanka Tasorinki (English translation: Lawrence Martin Schriewer) copyright Editorial PIKI E.I.R.L. First edition 2010.

INDEX

CPSIA information can be obtained
at www.ICGtesting.com
Printed in the USA
LVOW02s1328120717
540923LV00001B/1/P